# GUIDELINES FOR USE OF
# Vapor Cloud Dispersion Models

By
STEVEN R. HANNA
and
PETER J. DRIVAS

For
## CENTER FOR CHEMICAL PROCESS SAFETY
of the
American Institute of Chemical Engineers
345 East 47th Street, New York, NY 10017

**Library of Congress Cataloging-in-Publication Data**

Hanna, Steven R.
   Guidelines for use of vapor cloud dispersion models.

   Bibliography: p.
   Includes index.
   1. Atmospheric diffusion—Mathematical models.
2. Hazardous substances—Environmental aspects—
Mathematical models.   3. Vapors—Mathematical models.
I. Drivas, Peter J., 1947–      II. Title
III. Vapor cloud dispersion modeling.
QC880.4.D44H35    1987       533'.63        87-1166
ISBN 0-8169-0403-0

This book is available at a special discount when ordered in bulk quantities.
For information, contact the Center for Chemical Process Safety at the address
given above.

**ISBN 0-8169-0403-0**

# Contents

# Acknowledgments

The authors appreciate the support and guidance of the CCPS committee, which includes the following individuals: Thomas Carmody and Sandy Schreiber (AIChE staff members), Gary Page, Chairman (American Cyanamid), Gene Lee (Air Products and Chemicals), William Hague (Allied-Signal Corp.), Doug Blewitt (Amoco Corp.), Johnnie Pearson (EPA), Rudolf Diener (Exxon Research and Engineering Co.), Jerry Schroy (Monsanto Co.), Ronald Lantzy (Rohm and Haas), Gib Jersey (Mobile Research and Development), James Moser (Shell Development Co.), and David McCready (Union Carbide Corp.).

The document was also reviewed in depth by Dr. James Fay (MIT) and Dr. Alan Eschenroeder (private consultant). The efficient typing and manuscript production by Ms. Alice Kashmanian of Sigma Research Corporation is greatly appreciated. Finally, we acknowledge the cooperation of the dozens of individual scientists and engineers who prepared useful discussions of their models, who took time to complete the questionnaires, and who sent us copies of relevant manuscripts.

Steven R. Hanna
Sigma Research Corporation
394 Lowell Street, Suite 12
Lexington, Massachusetts 02173

# Nomenclature

| | |
|---|---|
| $A$ | Area of opening ($m^2$) |
| $A_p$ | Area of evaporating pool of liquid ($m^2$) |
| $B_e$ | Effective half width of cloud in DEGADIS (m) |
| $c$ | Specific heat (joules $kg^{-1}\,{}^{o}K^{-1}$) |
| $c_o$ | Coefficient of discharge ($c_o \sim 1$) |
| $C$ | Concentration ($kg\ m^{-3}$) |
| $C_c$ | Centerline concentration ($kg\ m^{-3}$) |
| $C_D$ | Drag coefficient (about $10^{-3}$) |
| $C'$ | Turbulent fluctuation in concentration ($kg\ m^{-3}$) |
| $C_L$ | Concentration limit ($kg\ m^{-3}$) |
| $C_{max}$ | Maximum ground level concentration ($kg\ m^{-3}$) |
| $C_p$ | Predicted concentration ($kg\ m^{-3}$) |
| $C_o$ | Observed concentration ($kg\ m^{-3}$) |
| $d$ | Effective pool diameter (m) |
| $d_p$ | Pipe diameter (m) |
| $d_t$ | Tank diameter (m) |
| $D$ | Diameter of release opening (m) |
| $D_m$ | Molecular diffusivity ($m^2\ s^{-1}$) |
| $E$ | Potential energy ($m^4\ s^{-2}$) |
| $E_o$ | Initial potential energy ($m^4\ s^{-2}$) |
| $f$ | Generalized function; pipe friction factor |
| $F_o$ | Buoyancy flux ($m^4\ s^{-3}$) |
| $Fr$ | Froude number |
| $g$ | Acceleration of gravity ($9.8\ m\ s^{-2}$) |
| $h$ | Depth of cloud or plume (m) |
| $h_c$ | Convection coefficient ($4.5\ W\ m^{-2}\,{}^{o}K^{-1}$); critical height (m) |
| $h_e$ | Transfer coefficient for evaporation ($1\ m\ s^{-1}$); effective cloud depth (m) |
| $h_o$ | Initial depth of cloud or plume (m) |
| $H$ | Height of liquid above puncture (m) |
| $H_a$ | Evaporation of aerosol (watts $m^{-2}$) |
| $H_c$ | Condensation of water vapor (watts $m^{-2}$) |
| $H_{ch}$ | Conduction heat flow into pool from ground (watts $m^{-2}$) |
| $H_e$ | Evaporation heat flux (watts $m^{-2}$) |
| $H_g$ | Ground heat flux (watts $m^{-2}$) |
| $H_s$ | Sensible heat flux (watts $m^{-2}$) |
| $I$ | Intermittency (fraction of non-zero concentrations) |
| $k$ | Thermal conductivity ($W\ m^{-1}\,{}^{o}K^{-1}$); normalized mass transfer coefficient |
| $k_g$ | Mass transfer coefficient ($m\ s^{-1}$) |
| $K$ | Eddy diffusivity ($m^2\ s^{-1}$) |
| $\underline{\underline{K}}$ | Eddy diffusivity tensor ($m^2\ s^{-1}$) |
| $K^c$ | Eddy diffusivity for concentration ($m^2\ s^{-1}$) |
| $K^m$ | Eddy diffusivity for momentum ($m^2\ s^{-1}$) |
| $K_y$ | Lateral eddy diffusivity ($m^2\ s^{-1}$) |
| $K_z$ | Vertical eddy diffusivity ($m^2\ s^{-1}$) |
| $K^\theta$ | Eddy diffusivity for enthalpy ($m^2\ s^{-1}$) |

**Nomenclature**

| | |
|---|---|
| $l$ | Scale length of cloud (m) |
| $L$ | Latent heat of vaporization (joules kg$^{-1}$) |
| $L_p$ | Length of pipe from break to first valve (m) |
| $L_t$ | Length of horizontal tank (m) |
| $m_a$ | Molecular weight of air (kg mole$^{-1}$) |
| $m_j$ | Molecular weight of component j (kg mole$^{-1}$) |
| $M$ | Total mass of gases in mixture (kg) |
| $M_j$ | Mass of component j (kg) |
| $M_o$ | Momentum flux (m$^4$ s$^{-2}$); initial mass of gas (kg) |
| $M_T{}^o$ | Initial mass of evaporable liquid (kg) |
| $n$ | Power law exponent for wind profile |
| $n_T$ | Total moles of liquid |
| $N_{Re}$ | Reynolds number |
| $N_{Sc}$ | Schmidt number |
| $N_{Sh}$ | Sherwood number |
| $p$ | Probability density functions; pressure (newtons m$^{-2}$) |
| $p_s$ | Saturation vapor pressure (newtons m$^{-2}$) |
| $P$ | Cumulative distribution function |
| $Q$ | Continuous source strength (mass per unit time) |
| $Q_f$ | Mass of liquid that flashes (mass) |
| $Q_i$ | Instantaneous source strength (mass) |
| $Q_l$ | Liquid mass emission rate (mass per unit time) |
| $r$ | Radius of pool (m); distance from cloud center to point (m) |
| $r_s$ | Distance from point of interest to monitoring site (m) |
| $R$ | Radius of cloud or plume (m); Gas constant (joules kg$^{-1}$ $^o$K$^{-1}$); Correlation coefficient |
| $R^*$ | Universal gas constant (8.31436 joules mole$^{-1}$ $^o$K$^{-1}$) |
| $R_o$ | Initial radius of cloud or plume (m) |
| $R_y$ | Lateral plume half width (m) |
| $Ri$ | Richardson number |
| $Ri_a$ | Local Richardson number |
| $Ri_b$ | Bulk Richardson number |
| $Ri_c$ | Critical Richardson number |
| $Ri_o$ | Initial Richardson number |
| $s$ | Distance along plume axis (m) |
| $S_r$ | Soil roughness factor |
| $t$ | Time (s) |
| $t_a$ | Averaging time (s) |
| $t_s$ | Sampling time (s) |
| $T$ | Temperature ($^o$K) |
| $T_a$ | Ambient temperature ($^o$K); averaging time (s) |
| $T_b$ | Normal boiling point of liquid ($^o$K) |
| $T_g$ | Temperature of gas ($^o$K) |
| $T_I$ | Integral time scale (s) |
| $T_p$ | Temperature of plume ($^o$K) |
| $T_s$ | Sampling time (s) |
| $u$ | Wind speed (m s$^{-1}$) |
| $\underline{u}$ | $\underline{i}u + \underline{j}v + \underline{k}w$  Vector wind speed (m s$^{-1}$) |

$u_e$      Effective velocity of cloud (m s$^{-1}$)

$u_s$      Plume speed in direction of plume axis (m s$^{-1}$); sonic velocity in pipe rupture (m s$^{-1}$)

$u_*$      Friction velocity (m s$^{-1}$)

$U_E$      Edge entrainment velocity (m s$^{-1}$)

$U_f$      Velocity of gravity front (m s$^{-1}$)

$U_T$      Top entrainment velocity (m s$^{-1}$)

$v_*$      Characteristic turbulent velocity (m s$^{-1}$)

$V$:      Volume of cloud (m$^3$)

$V'$      Volume flux of continuous plume (m$^3$ s$^{-1}$)

$V_o$ and $V_o'$      Initial values

$V_1$      Liquid volume remaining in tank (m$^3$)

$w$      Vertical speed (m s$^{-1}$)

$w_e$      Entrainment velocity (m s$^{-1}$)

$w_o$      Initial plume vertical speed (m s$^{-1}$)

$w_*$      Convective velocity due to ground heating of cloud (m s$^{-1}$)

$x$      Along-wind distance (m)

$x_o$      Initial vapor mass fraction

$x_f$      Distance where gravity slumping ceases (m)

$x_g$      Distance to plume touchdown (m)

$X$      Soil roughness factor

$y$      Lateral distance from plume center (m)

$y_o$      Lateral position of plume center (m)

$z$      Vertical distance above ground (m)

$z_e$      Height of plume center above ground (m)

$z_h$      Hill height (m)

$z_c$, $z_r$      Reference heights in Wilson (1981) model (m)

$z_o$      Roughness length (m)


$\alpha$      Alignment weighting, power law in DEGADIS model

$\alpha_s$      Thermal diffusivity of soil (m$^2$ s$^{-1}$)

$\beta$      Constant in Wilson (1981) model(s)

$\gamma$      Gas specific heat ratio

$\delta$      Dirac delta function, pipe flow constant

$\Delta C$      Error in concentration due to measurement errors (kg m$^{-3}$)

$\Delta h$      Plume rise (m)

$\overline{\nabla}$      Del operator $\underline{i}\,\partial/\partial x + \underline{j}\,\partial/\partial y + \underline{k}\,\partial/\partial z$

$\epsilon$      Eddy dissipation rate (m$^2$ s$^{-3}$); small perturbation

$\upsilon$      Kinematic viscosity (m$^2$ s$^{-1}$)

$\omega$      Correlating parameter for flashing (see eq. 4-7)

$\rho_a$      Ambient density (kg m$^{-3}$)

$\rho_{aer}$      Aerosol density per unit volume of cloud (kg m$^{-3}$)

$\rho_g$      Gas density (kg m$^{-3}$)

$\rho_{gs}$      Saturated value of gas density (kg m$^{-3}$)

$\rho_1$      Liquid density (kg m$^{-3}$)

$\rho_o$      Cloud density (kg m$^{-3}$)

$\rho_p$      Plume density (kg m$^{-3}$)

$\sigma_c$      Standard deviation of concentration fluctuations (kg/m$^3$)

**Nomenclature**

| | |
|---|---|
| $\sigma_r$ | Horizontal dispersion parameter (m) |
| $\sigma_x$ | Along-wind dispersion parameter (m) |
| $\sigma_y$ | Lateral dispersion parameter (m) |
| $\sigma_z$ | Vertical dispersion parameter (m) |
| $\theta$ | Angle of plume axis with horizontal (radians); angle of liquid surface in tank (radians) |
| $\phi$ | Dimensionless function of Ri |
| $\phi_s$ | Angle between observed wind direction and the line from the monitoring site to the point of interest (radians) |

# 1

# Introduction

<u>BACKGROUND</u>

The American Institute of Chemical Engineers (AIChE) has a 30-year history of involvement with process safety and loss control for chemical and petrochemical plants. Through its strong ties with process designers, builders, and operators, safety professionals, and academia, the AIChE has enhanced communiction and fostered improvement in the high safety standards of the industry. Their publications and symposia have become an information resource for the chemical engineering profession on the causes of accidents and means of prevention.

Early in 1985 the AIChE established the **Center for Chemical Process Safety** (CCPS) to serve as a focus for a continuing program for process safety. The first CCPS project was the publication of a document entitled <u>Guidelines for Hazard Evaluation Procedures</u>. The second CCPS project is the present document.

<u>PURPOSE</u>

Modeling episodic releases of hazardous or toxic materials is a rapidly evolving field in chemical engineering, driven by the efforts of the industry to prevent and mitigate such incidents, as well as by public concern beginning to be reflected in regulatory requirements. <u>Guidelines for Use of Vapor Cloud Dispersion Models</u> is intended to assist this effort by providing in one publication an overview of the subject that will be of value to the beginner, the expert, and the manager- whether in industry or a regulatory agency. For the beginner, it is a primer- or starting point- which we hope will provide perspective and facilitate rapid progress into the field of dispersion modeling. For experts, it is an overview document and a resource, perhaps useful as a text for those apprenticing under their care. For managers, it is a bridge linking to the expert, permitting the manager to finally understand the difficulties, uncertainties and limitations associated with modeling episodic releases. This bridge is equally important in both industry and the regulatory agencies which may become involved. Finally, we hope that it may also be useful in academia.

Overall, the objective of <u>Guidelines for Use of Vapor Cloud Dispersion</u>
<u>Models</u> is to help facilitate the development and use of dispersion modeling as
an everyday tool within the industry, along with an understanding of the
limitations of that tool.  It is recommended to the beginner and the expert as
a primer and resource and particularly to the manager for a quick review—
whether in industry or a regulatory agency.

<u>COMMENTS</u>

Much of the information used to prepare this book was obtained through the
generosity of individual researchers.  Since the field of hazardous gas source
emissions and transport/dispersion of hazardous chemicals is rapidly evolving,
an attempt was made to obtain the most recent reports and manuscripts from
model developers.  A survey of the current literature on this subject shows
that the person who has a practical need for such a model can easily become
confused.  This person needs or requires two things:

1.  A practical understanding of the basic physical and chemical
    principles.

2.  Information to permit him to choose the best available model
    for his situation.

This document purposely presents this material in a concise way so that it can
be easily used.

The description of practical models in Appendix A is based on the results
of questionnaires sent to model developers.  Rather than rely on third-person
reviews of models, the developers were requested to describe the capabilities
of their models.  There is a risk that some modelers will exaggerate the
capabilities of their models, but this shortcoming is outweighed by the benefit
that the modeler cannot claim that his model was misrepresented.

# 1. Introduction

Most of the examples of special applications in Appendix B were
contributed by members of the CCPS committee, who are in touch with realistic
scenarios of accidental releases of hazardous materials to the atmosphere.  It
is hoped that these examples will aid the reader in understanding the types of
scenarios that he may encounter.

The body of the text is broken into five sections:

- Overview of Problems and Modeling Procedures
- Input Data
- Source Emission Models
- Transport and Dispersion Models
- Model Evaluation and Uncertainty

The philosophy of writing is to first explain the basic physical
principles of a phenomenon and then describe a few examples of related models
or research programs.  No attempt is made to provide a comprehensive review of
all works on this subject, and in many cases the only place a particular model
may be mentioned is in a summary table.  This procedure must be followed in
order to produce a book that is concise enough to be usable.  The models are
found to be sufficiently similar that many of them can be covered by a single
description.  For example, the SPILLS source emission algorithm is used in many
other models.

Gaps in our knowledge are also pointed out in the text.  For example,
there are no sufficiently accurate models for two-phase jets contained in the
various microcomputer-based hazard response models, even though it is now
recognized that many jets of interest are two-phase.  This review document,
therefore, represents a snapshot of the state-of-the-art in December, 1986, of
a very rapidly-developing field.  Within one year, many models will be
improved, more models will appear on the scene, and more field experiments will
be conducted.

# 2

# Overview of Release Scenarios
# and Modeling Procedures

A few examples of catastrophic releases of hazardous chemicals have been heavily reported by the news media, but most releases that the engineer must deal with on a day-to-day basis are relatively minor and less life-threatening. It is stressed that the same models are used to treat the large and small scenarios, since the basic physical principles are independent of the size of the release. Table 2-1 contains a brief summary of some hypothetical source scenarios, and more details are given in Appendix B. These examples can be grouped into three broad areas:

(1) Tank rupture, with subsequent spill and evaporation.
(2) Pipe break, with initial high velocity jet.
(3) Venting of runaway reaction.

Similarly, there are a variety of environmental impact questions that must be taken into account, such as the location, magnitude, and duration of a certain air concentration. There may be toxicological and flammability information that would strongly influence the averaging time used by the model. The engineer should carefully check these data for the chemical in question before proceeding with any model calculations.

The choice of a source emissions model and transport and dispersion model is dependent on the chemical, the release scenario, and the desired averaging time. Chapters 4 and 5 provide detailed summaries of the physical principles underlying these models. The major regions of hazardous gas modeling in the atmosphere are drawn in Figure 2-1, showing how the initial source emissions and acceleration phases quickly give way to a regime in which the internal buoyancy of the plume or puff dominates the dispersion. This regime is followed by a transition to a regime in which ambient turbulence dominates the dispersion. For some questions, such as the lower flammability limit (LFL) of LNG, the calculations need not go beyond the gravity slumping (dominance of internal buoyancy) regime, since the LFL occurs at a relatively high concentration of a few percent. However, for questions such as the one hour average concentration at a subdivision located 10 km from the release, the

Table 2-1

Examples of Hypothetical Hazardous Chemical Release Scenarios

| Chemical and Container | Release Type |
|---|---|
| Chlorine, one ton tank, 500 psig | (1) Tank rupture under fire<br>(2) Spill due to crack while handling<br>(3) Tubing broken (0.5 in. dia, 10 ft. length)<br>(4) 0.5 in. safety plug pops |
| Chlorine, 2 in. dia. pipeline | Broken pipeline |
| Ammonia pressurized storage tank, 40 metric tonnes, $20^{\circ}C$, 7.5 bar (g) pressure | (1) Overfill, spill at liquid fill rate<br>(2) Rupture of 1.25" line, spill on undiked concrete |
| Hydrogen sulfide still | Failure results in release from 85 ft. vent of 160 lb in 1 minute at $55^{\circ}C$ |
| Hydrogen sulfide pipeline (10% mass conc. in nat. gas) | 12" buried pipeline severed,, $T=300^{\circ}K$, p = 1000 psig |
| HCH tank, 12 ft. head | Valve broken, $54^{\circ}F$ |
| Trimethoxysilane tank | Safety release valve (0.33 ft. dia) rupture, 1898 g/s, $412^{\circ}K$, vert. release at 42 ft. height |
| Acrolein tank car | 2" hose break, 200 gpm |
| Butadiene gas and water | Runaway reaction in 7000 gal. emulsion kettle, 150 mm dia. vent line, 25m height |
| Hydrogen cyanide | Runaway reaction, relief valve, 35000 lb/hr. |

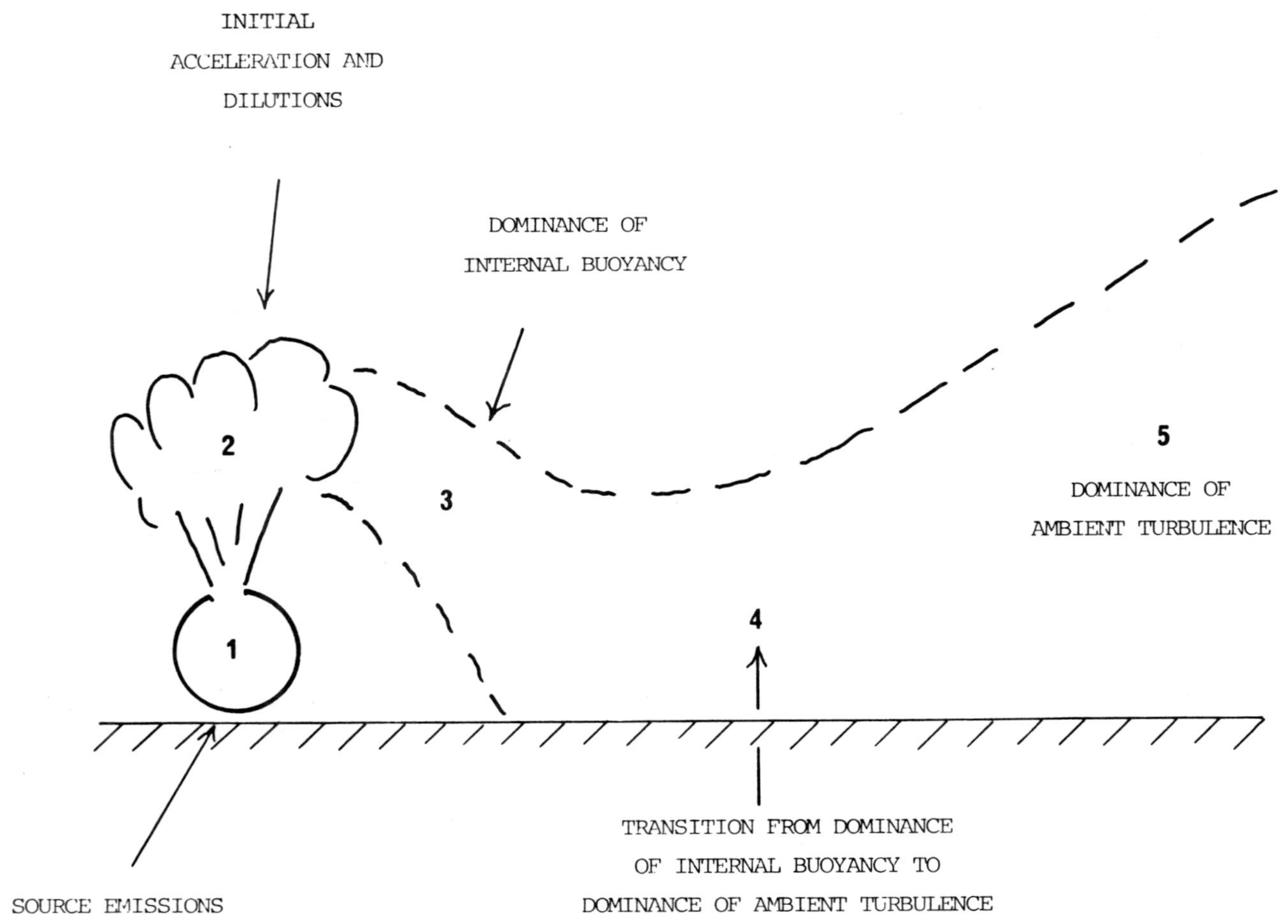

Figure 2-1. Illustration of five major regions of hazardous gas modeling.

calculations must extend well into the regime where ambient turbulence is dominant.

In modeling vapor cloud dispersion, there are a number of steps that must be followed, with many pathways and options. A figure containing all the steps and options would be much too long and confusing to present here. Figure 2-2 is a one-page summary of the general sequence of logic that should be followed. It also serves as an outline for the remainder of the discussions in this book. It is stressed that there are currently no modeling systems available which cover all the items in Figure 2-2. For example, many models treat dense gas slumping while no models treat two-phase jets or aerosol formation. Also, in no practical cases are all of the required input data observed.

In any particular scenario, only the relevant parts of the sequence in the figure would be followed. For example, for a gas jet, it is not necessary to know the environmental properties of the underlying surface and only the gas jet source emission model need be used.

The comparison in Table 2-2 is given to stress the inherent difficulties in modeling accidental releases, as opposed to modeling routine air pollutant releases (e.g., $SO_2$ released from a power plant stack). The differences between the two sets of conditions clearly indicate that the relatively simple Gaussian plume models which have been used for many years for stack gas dispersion are typically not appropriate for modeling accidental releases. The Environmental Protection Agency (1984) recognizes these differences but emphasizes that models are not available or tested for many scenarios. The remainder of this Guideline describes models which are appropriate for these types of releases. None of these models, however, have been approved by the EPA.

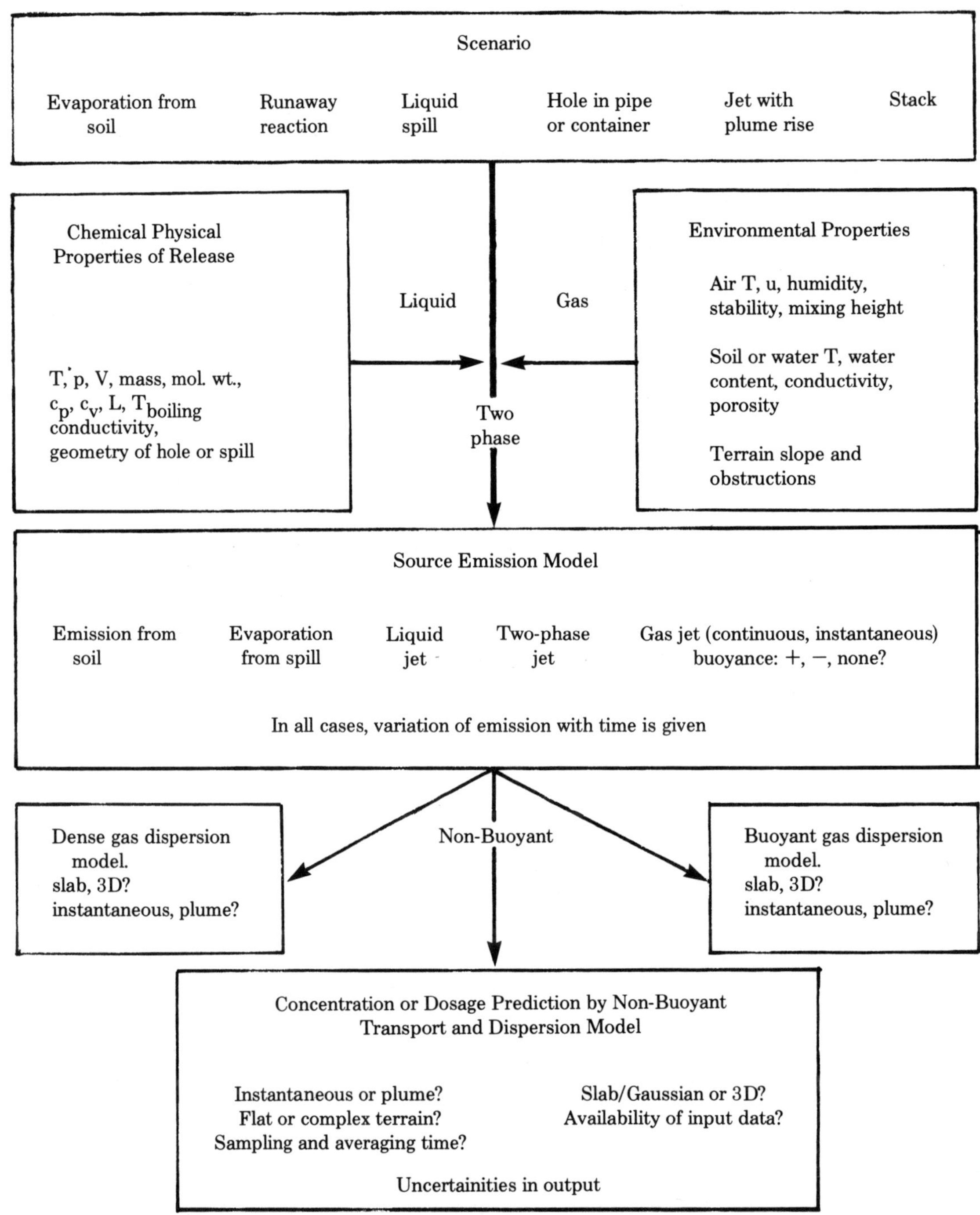


Figure 2 2. Generalized Vapor Cloud Model Logic Sequence
In any specific case, only a portion of the sequence would be followed.

Table 2-2

Comparison of Characteristics of Emissions of
Routine versus Accidental Releases

Routine Release                          Accidental Release


- Source well-defined                    - Source poorly-defined
  (physical and chemical parameters        (physical and chemical parameters
  easily measured).                        not all measured).


- All gas release.                       - Gas and/or liquid release.


- Continuous, steady-state               - Usually highly transient.


- Little thermodynamic interaction       - Often strong thermodynamic
  with ground/water surface.               interaction with surface.


- No phase changes.                      - Frequent phase changes.


- Generally one hour averaging           - Variety of averaging times,
  time.                                    from one second to one hour or
                                           longer


- No dense gas effects.                  - Possible dense gas slumping,
                                           terrain-following.

# *3*

# **Input Data Required for Modeling**

Source emissions, meteorological conditions, and terrain can be highly variable and must be well-known in order to assure optimum performance by any model.  If the concentration is directly proportional to an input parameter, then a 10% uncertainty in that parameter leads to a 10% uncertainty in the predicted concentration.  The discussions below provide an overview of the types of input data required by the models, and Section 6 will provide more details on the mathematical aspects of uncertainty analysis.

3.1   Source Data

This book contains a separate section on source emission modeling, which covers, for example, estimation of the evaporation of a spilled liquid.  But even the source emissions model requires several pieces of input information in order to operate, as discussed in the previous section on source scenarios. The most important of these is the total mass of the release (if instantaneous) or the mass release rate (if distributed over time).  In most models, the error in concentration prediction is directly proportional to the error in mass release rate.  The required source data are listed below:

a.   <u>Physical and Chemical Characteristic of Material</u>.  First it is necessary to know the chemical composition of the liquid and/or gas, permitting the molecular weight to be determined.  The molecular diffusivity, conductivity, boiling point, and other characteristics of the material can then be found from standard references.  Physical characteristics of the materials are also required, such as vapor pressure correlations with temperature.  If there are several components in a mixture, the characteristics of each component must be known.

b.   <u>Geometry of Source</u>.  The release rate is related to the geometry of the source, which includes both the dimensions of the pipe, tank, or stack, and the characteristics of the source orifice.  These parameters permit estimation of the release amount and type (e.g. a burst tank results in an instantaneous release, while a small hole in a tank results in a continuous release). The position of the orifice relative to the level of the liquid can determine whether the release is gaseous or liquid or a combination of both. There are clearly several types of source geometries, ranging from thin pipes to large tankers. It is also important to know the location of the source above the ground or whether part of the source is buried. If a pipe is involved, the direction of discharge should be determined.

c.  <u>Knowledge of Plant Operating Procedures</u>.  Most plants have special safety procedures that influence the source release rate. For example, a safety valve may be on a pressurized tank to prevent the tank from rupturing in a fire. Or a gas pipeline may have valves which automatically close in the event of a rupture. Each of these systems may limit the amount and duration of the release.

d.  <u>Time Variation</u>.  Maximum predicted concentrations are likely to be proportional to the maximum source release rate. If the release rate is highly variable in time, then along-wind dispersion must be accounted for in the transport and dispersion model. Consequently the time variability of the source is required. The flow rate from most pressurized containers follows an exponential formula, with a certain time scale that must be estimated. (Wilson 1981)

e.  <u>Characteristics of Underlying Ground Surface</u>.  Many hazardous releases are cold or dense gases or liquids which flow on the ground and exchange heat and moisture with the ground. At some chemical plants, dikes are built in order to contain spilled liquids. If there is no dike, the fluid will flow in an uncontrolled manner and evaporative emission rate will be larger. Clearly it is important to know the thermal conductivity, density and specific heat of the ground, the moisture content, the porosity and organic carbon content of the soil and the terrain contours.

f.   <u>Presence of Gas, Liquid, and Aerosol</u>. No practical source emissions models can directly estimate the fraction of gas, liquid, and aerosol (i.e., two phase cloud) in a release, but most require these parameters as input. It is emphasized that this is one of the great uncertainties in modeling. This is an active area of research but there are many difficulties in collecting observations of two-phase jets and clouds.

## 3.2 Meteorological Data

Modeling can be performed for either real-time input data or for historical or hypothetical conditions. Because of the operational difficulties and uncertainties associated with real-time modeling of hazardous gases, most modeling exercises now use historical or hypothetical situations for planning purposes. In the latter case, it is not so important to calculate the plume or cloud trajectory based on wind direction observations. Standard EPA models require a measurement of the wind velocity at a height of 10m and an indication of the cloudiness. From this information and a knowledge of the hour of the day and the time of the year, the model can estimate the velocity and dilution rate of the gas plume or cloud. More recent models can more accurately treat dilution by using observations of turbulence intensity. As shown in Section 5, the dilution rate of dense gas clouds is initially independent of atmospheric turbulence and the only meteorological parameter that is required for concentration predictions at short ranges is the wind velocity.

Winds are quite variable in space. Typical root-mean-square errors in hourly averaged wind direction are $30^{\circ}$ for two wind vanes spaced 10 km apart in flat terrain (Hanna 1982). This underlines the importance of having <u>on-site</u> meteorological measurements. If wind data are obtained from a National Weather Service installation at a nearby airport, it must be recognized that the data will not be representative. However, it is also important to locate the on-site instruments in open areas at distances at least ten building heights away from buildings and other structures. It should be recognized that NWS data are not even representative of hourly averages at the site, since they are read only to the nearest $10^{\circ}$ and are averaged for only 30 seconds before and after the hour.

If the plant is located in a region where there may be complex flow
patterns, such as near a shoreline or in a region with mountains and valleys,
it may be necessary to set up several wind measurement sites.  In this manner,
typical flow patterns can be determined (although very few models can account
for this complexity).  Such wind monitoring networks are now in place in many
government facilities which routinely handle hazardous materials (e.g. Savannah
River Laboratory, Dugway Proving Ground).

Stability can be calculated from the wind speed measurements and from an
estimation of the net radiation flux.  This procedure is usually carried out by
a meteorological preprocessor algorithm within the model.  Otherwise,
procedures outline in Section 5.1.3 can be used.  It is stressed that a given
stability class is maintained only for a few hours, at most, due to the
influence of the solar cycle on the boundary layer.

If the hazardous gas is cold or contains aerosols, it is possible that
entrained ambient water vapor will condense within the cloud or that existing
aerosols will evaporate.  Observations of ambient relative humidity are
required to calculate this effect.

3.3  Site Information

If the hazardous material is a dense gas, its transport and dispersion are
highly dependent on local topography, including both natural and man-made
features.  The dense gas flows like water across the site, seeking low spots
and deflecting around hills and buildings.  These features also affect inert,
neutrally-buoyant gases, although in this case their primary influence is upon
short-range dilution or dispersion.  It is generally assumed that a non-buoyant
gas released in the wake of a building of a given cross-sectional area will
initially be diluted uniformly across that area.  Unfortunately most models
neglect these two aspects.

For model applications it is clear that a map of the site with topography
and building dimensions must be included in the set of input data.  Some
analysis may be required to determine the initial dilution in the lee of the

building for releases at different locations for different wind directions, or for estimating the path of the dense gas flow for these conditions.

Some models require information on site roughness. The surface roughness parameter, $z_o$, which equals about one-tenth of the height of the roughness elements, is about 1 m for cities, forests, and industrial complexes, 1Ø cm for agricultural crops, 1 cm for grass, and 1 mm for water or pavement surfaces. If the roughness elements are higher than the diffusing cloud, as they may be for a very dense cloud in a forest, then most models are inapplicable. Another special case occurs when the site contains numerous buildings, pipelines, tanks, and other obstructions to the flow. In some cases it is necessary to use wind tunnels or water channels to determine the transport and dispersion of a plume released within these obstructions.

## 3.4 Receptor Data

It may seem obvious that the location of important receptors such as monitoring sites, population centers, sensitive crops, and property boundaries be considered part of the input data. But the reason for placing this subject into a separate subsection is to place emphasis on the roles of sampling and

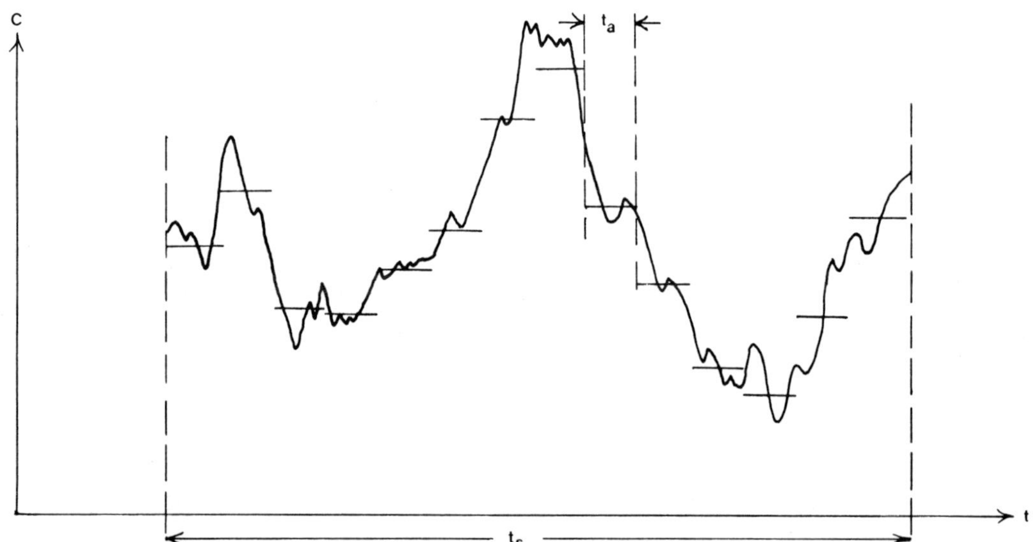

Figure 3.1  Illustration of Concentration Time Series with Sampling Time, $t_s$, and Averaging Time, $t_a$, Indicated.

averaging times, which are often ignored in hazardous material studies. The sampling time, $t_s$, is the total duration of time that a monitor is operating or that a modeled plume is assumed to impact a receptor location. The averaging time, $t_a$, is the length of time over which data are averaged to produce the concentration time series. This averaging time may be imposed by a monitoring instrument or a biological process. For example, instantaneous averaging times are of interest for a flammable gas, while one hour averaging times may be appropriate for low concentrations of a toxic gas such as $H_2S$. The difference between $t_a$ and $t_s$ is shown in the example in Figure 3-1. For a given sampling period, the maximum concentration during the period increases as averaging time $t_a$ decreases.

Averaging time is also important for dispersion from instantaneous sources, which may impact a monitor for only a short period of time. In this case the dosage (integral or sum of concentration over time) is constant but the average concentration decreases as $t_a$ increases.

Sometimes averaging distance, area, or volume must also be specified. For example, if a single breath is important, then the representative averaging volume is about .001 $m^3$. For flammable gases, the instantaneous concentrations are important, and the location of potential ignition sources must be known. If the average impact on the population in a city block is important then the averaging area may be about 1 $km^2$. This information is tied to toxicological data for that particular hazardous gas.

After all this discussion on the importance of sampling and averaging times, it must be emphasized that very few models can account for variations in these times. Usually these times are implicitly built into the empirical model with little or no discussion.

Finally, the important question of indoor air pollution often arises. Knowing the air exchange rate of a building, indoor concentrations can be predicted based on the time series of outdoor concentrations. A few preliminary models exist for this phenomenon.

A summary of the input data requirements discussed in this section is given in Table 3-2. Any given model only will require a few of these input data.

Table 3-2
Summary of Input Data Requirements
(Not all are Required in any Given Model)

Source Data

Physical and Chemical Characteristics (molecular weight, diffusivity,
viscosity, conductivity, vapor pressure, temperature, boiling point,
specific heat, latent heat, vapor-liquid equilibria data).

Geometry of Source (dimensions of facility and release hole, position of
release relative to liquid or gas in tank, height above ground).

Plant Operating Procedures (characteristics of safety valves and plugs,
control valve closing times, ventilation devices).

Time Variation of Release Rate.

Underlying Ground Surface (topography, dike position, thermal
conductivity, density, specific heat, moisture content, and porosity
of ground).

Fraction of Gas, Liquid, and Aerosol.

Meteorological Data

On-site wind velocity (one representative site if terrain is flat,
multiple sites if terrain is complex).

On-site net radiation flux (default: cloudiness).

Temperature, relative humidity, turbulence, stability.

Site Information

Local topography, including building and storage tank dimensions, dike
dimensions, equipment and operating information.

Receptor Information

Location, sampling and averaging times, spatial averaging over line,
area, or volume.

Air exchange rate for a building.

Toxicological data; location of ignition sources

# 4

# Source Emission Models

Modeling the source phenomenon for an accidental spill of hazardous material is perhaps the most critical step in the accurate estimation of downwind concentrations. Because of the nature of accidental spills, it is possible that some key model parameters, for example the volume of spilled material, can be only roughly estimated. Any inaccuracies in the source emission estimation will greatly influence the subsequent dispersion calculation of concentrations resulting from an accidental release. For example, if the source were treated as an instantaneous explosion, the predicted concentrations would be quite different than if the source were treated as a continuous area source. The dimensions of the source emission term depend on the scenario. For example, this term would have units of $g\ m^{-2}s^{-1}$ for continuous evaporative emissions from a liquid spill, or g for an instantaneous release. It is recognized that no phenomenon in the atmosphere is truly "instantaneous," but for modeling purposes, it is convenient to represent releases with a duration of a few seconds as instantaneous.

This section is divided into two main parts. Section 4.1 describes the basic physical and chemical principles that are appropriate in various types of spill emission situations, to enable an engineer to identify and understand the basic techniques used for calculating the source emissions from a particular accidental release. It is stressed that many liquid spills are multicomponent and many emissions from tank or pipe ruptures are two phase, and prediction schemes for these scenarios are not well-developed or tested. A model for binary components does exist (Wu and Schroy 1979) and has been tested with limited data. Section 4.2 presents a review of specific emission calculation techniques that are used in some current source emission models.

4.1  Discussion of Physical and Chemical Principles

Accidental releases of hazardous materials can be of many different types - - gas or liquid, instantaneous or continuous, from storage tanks or pipelines, refrigerated or pressurized, on land or water, confined or unconfined, reactive or non-reactive. In many cases, combinations of these scenarios exist simultaneously.  The purpose of this section is to describe the important physical processes in various types of releases, and to present the general concepts that can be used to calculate the emission rates from these releases. Where applicable, basic equations will be presented from standard references such as Perry's Chemical Engineers' Handbook (Perry et al., 1984, revised on a regular basis).

Storage tank failures can result from corrosion, thermal fatigue, inlet or outlet pipe rupture, or valve failure.  Figure 4-1, reproduced from Fryer and Kaiser (1979), shows some of the different release mechanism scenarios that can occur with pressurized tanks and refrigerated liquids.  Our current knowledge (e.g. Fauske, 1985) suggests that "most emergency releases involve two-phase flows," and thus wherever vapor or liquid jets are indicated on the figure, the actual release could be vapor and liquid.  Obviously, there will be a large difference in the time variability of emissions and method of calculation for a catastrophic tank failure (instantaneous release) in comparison with a small puncture failure in a storage tank (continuous release).  Also, different calculation techniques may apply depending on whether a tank failure occurs in the liquid region or in the vapor space above the liquid, and whether the release is one or two phase.

Other potential release scenarios were listed in Table 2-2 and will be discussed in more detail in Appendix B.  For example, in addition to the tank failures discussed above, there are potential problems with runaway reactors, pipeline breaks, and explosions.

The following subsections will review the physical processes relevant for several of these emissions scenarios.

4.1.1  Gas Jet Releases

One important type of accidental release is a gas jet from a small puncture in a pressurized pure gas pipeline or in the vapor space of a pressurized gas storage tank, as shown in the top left drawing of Figure 4-1. Intuitively, one would expect to see an initial high release rate that

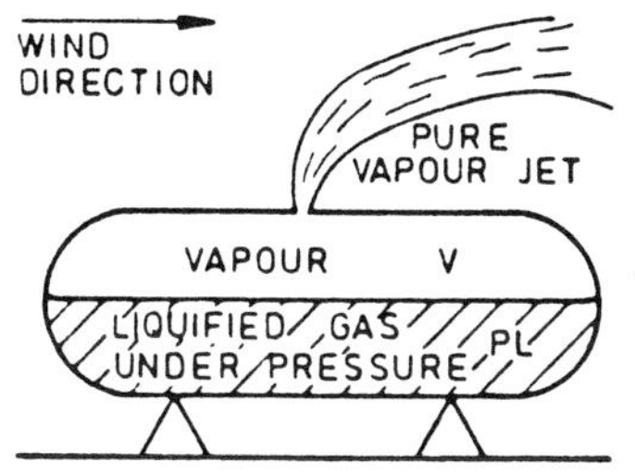



FIG 6A SMALL HOLE IN VAPOUR SPACE-PRESSURIZED TANK

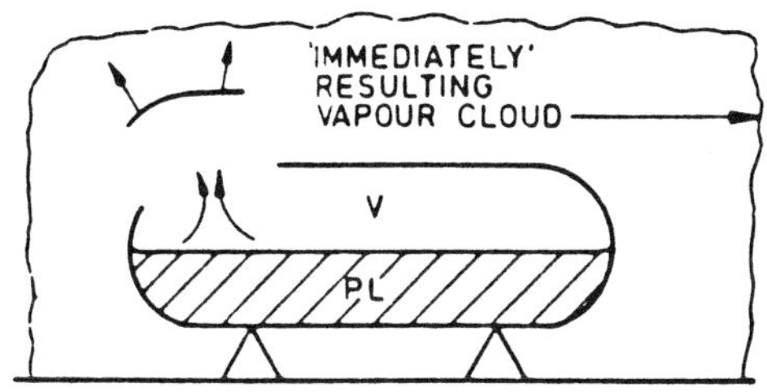



FIG 6B CATASTROPHIC FAILURE OF PRESSURIZED TANK

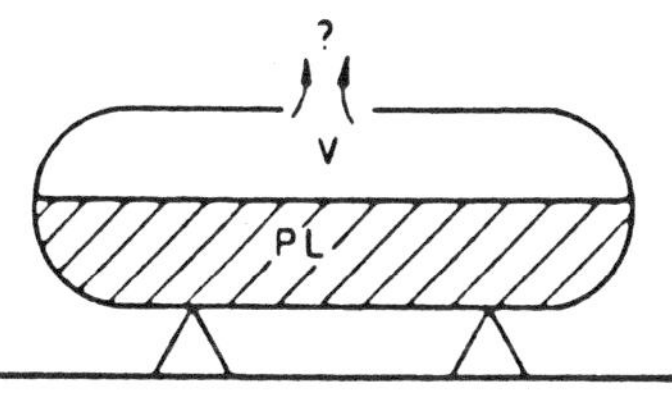



FIG 6C INTERMEDIATE HOLE IN VAPOUR SPACE-PRESSURIZED TANK

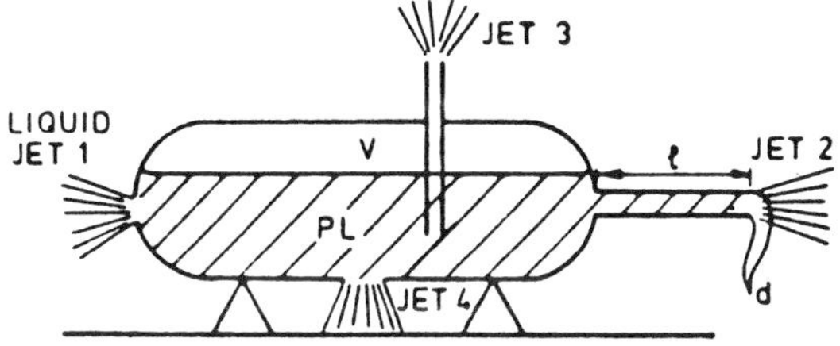



FIG. 6D ESCAPE OF LIQUIFIED GAS FROM A PRESSURIZED TANK

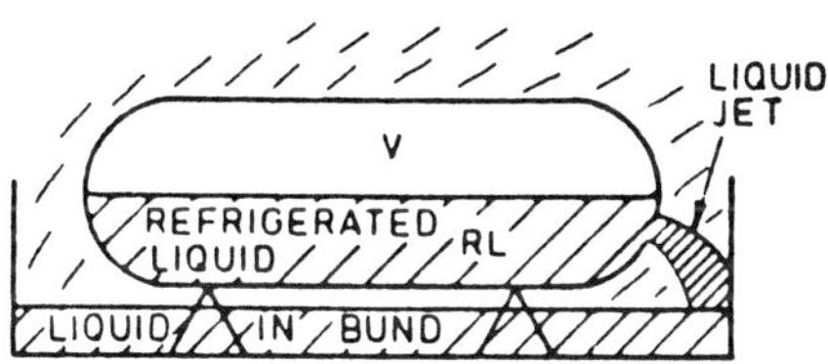



FIG 6E SPILLAGE OF REFRIGERATED LIQUID INTO BUND

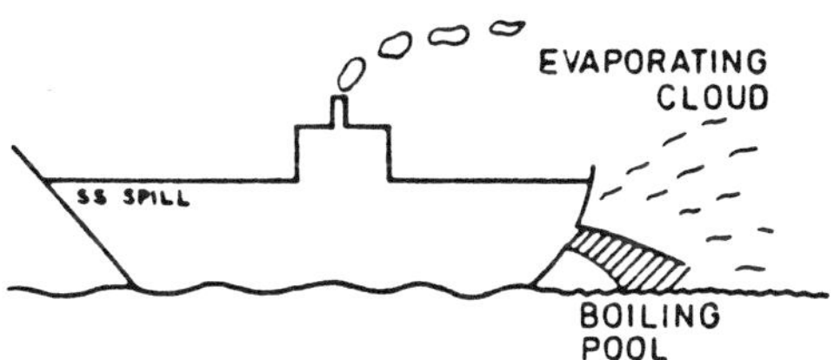



FIG. 6F SPILLAGE OF REFRIGERATED LIQUID ONTO WATER

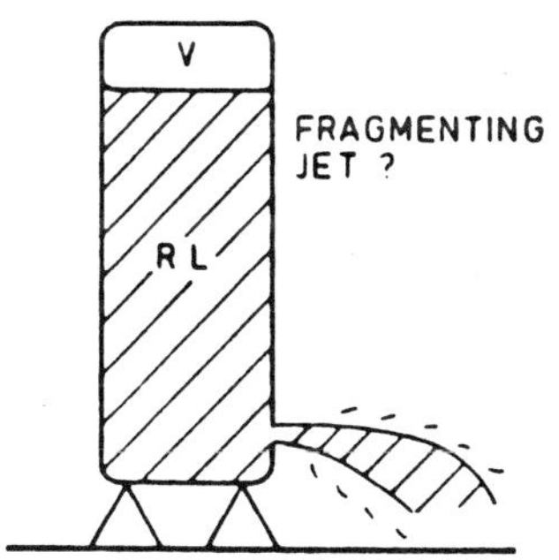



FIG 6G HIGH VELOCITY FRAGMENTING JET FROM REFRIGERATED CONTAINMENT

Figure 4-1.   Illustration of some conceivable release mechanisms, from Fryer and Kaiser (1979).  In most of these cases, the jet could be two phase (vapor plus entrained liquid aerosol).

decreases as the gas tank pressure decreases.  When a small puncture occurs, gas can exit the puncture only as fast as its sonic velocity, because of a "choked" condition at the exit.  Choked flow or critical flow simply means that the gas is moving through the puncture at its maximum possible speed, namely the speed of sound in the gas.

The pressure criterion for choked or critical flow to occur, assuming an ideal gas exiting through a small hole, can be simply expressed (Perry et al., 1984):

$$p/p_a \geq ((\gamma+1)/2)^{\gamma/(\gamma-1)} \tag{4-1}$$

where $p$ = absolute tank pressure (newtons m$^{-2}$)

$p_a$ = absolute atmospheric pressure (newtons m$^{-2}$)

$\gamma$ = gas specific heat ratio (the heat capacity at constant pressure, $c_p$, divided by the heat capacity at constant volume, $c_v$)

For a typical gas value of $\gamma = 1.5$, this critical $p/p_a$ ratio, using equation (4-1), is approximately 2.  Thus, if the tank pressure remains above approximately twice atmospheric pressure, the gas release rate will remain choked or critical.  For an ideal gas exiting through an orifice under isentropic conditions, the gas emission rate will follow the critical flow relationship (Perry et al., 1984):

$$Q = c_o Ap((\gamma m/R^*T)(2/(\gamma+1))^{(\gamma+1)/(\gamma-1)})^{1/2} \tag{4-2}$$

where $Q$ = gas mass emission rate (kg s$^{-1}$)

$c_o$ = discharge coefficient for orifice (dimensionless)

$A$ = puncture area (m$^2$)

$m$ = gas molecular weight (kg mole$^{-1}$)

$R^*$ = gas constant (8.31 joules mole$^{-1}$ $^oK^{-1}$)

$T$ = absolute gas temperature in the tank ($^oK$)

The coefficient of discharge (usually slightly less than 1) accounts for reductions below the theoretical velocity due to viscosity and secondary flow effects.  It depends upon nozzle shape and Reynolds number.  For a conservative or maximum gas release rate, $c_0 = 1$ can be used in equation (4-2).  When the tank pressure decreases to the criterion in equation (4-1), or below about twice atmospheric pressure, choked flow no longer applies and the flow rate becomes subcritical (Perry et al., 1984):

$$Q = c_0 A \{2 \rho_p p (\gamma/(\gamma-1)) ((p_a/p)^{2/\gamma} - (p_a/p)^{(\gamma+1)/\gamma})\}^{1/2} \qquad (4-3)$$

where $\rho_p$ = gas density in tank (kg m$^{-3}$).

Although the flow through the puncture remains sonic until the pressure drops to the criterion in equation (4-1), it should be noted from equation (4-2) that the mass flow rate exiting will steadily decrease.  This is because the pressure of the gas in the pipe or tank is decreasing with time.  One will see initially very high mass flow rates that rapidly diminish with time in a roughly exponential decay manner.

To calculate the pressure to use in equations (4-2) or (4-3), an assumption must be made whether the gas release occurs isothermally or adiabatically.  A pipeline release can be isothermal in cases with relatively small releases, because the expansion cooling is counteracted by frictional heating and heat transfer through the pipe walls.  In other cases involving more rapid gas releases or insulated pipes, adiabatic conditions may be more appropriate.  Wilson (1979) presents detailed equations for both the isothermal and adiabatic conditions in gas pipeline failures.  Bell (1978) has developed a simple empirical correlation for a pipeline release that compares very well with the more detailed calculations of Wilson (1979).

For the case of a gas jet exiting from a pressurized or cryogenic liquid storage tank, adiabatic conditions can be usually assumed for insulated storage tanks.  If a polytropic process ($p\rho^{-n}$ = constant) is assumed, then the adiabatic case is given by $n = k = c_p/c_v$ and the isothermal case is given by $n=1$, where in this case $p$ and $\rho$ are the pressure and density in the tank.

Because the gas release is continually supplied by liquid evaporating (in the case of cryogenic liquids, boiling) in the tank, the heat of vaporization of the liquid must be taken into account to calculate the temperature and pressure to use in equation (4-2). For an adiabatic system, the heat of vaporization is supplied primarily from the heat capacity of the liquid, and to a lesser extent from the heat capacity of the gas phase and the storage tank. Thus, in an insulated storage tank release, the liquid phase will become continually colder as more liquid evaporates. If the release takes place over a relatively long time period, heat transfer into the tank must also be considered. When all the liquid has evaporated, the release can be treated as a pure gas failure case.

### 4.1.2 Liquid Jet Releases

A puncture or failure in the liquid space of a pressurized or cryogenic storage tank is considerably more complicated than the corresponding case of a vapor space failure. In the liquid release case, a liquid jet is typically propelled from the tank and, depending on the normal boiling point of the compound or the flashing behavior of a multicomponent mixture, it may instantaneously vaporize or "flash" a large portion of the liquid. Figure 4-1 shows some of the possible types of liquid jet releases. In this subsection, the simplified case of a pure liquid release is reviewed.

Intuitively, one would expect that the release rate of a liquid jet would be proportional to the amount of pressure on the liquid in the tank, and would also depend on gravity because of the height of the liquid above the puncture. The equation typically used for calculating the liquid flow rate through a small puncture is based on the classical work of Bernoulli and Torricelli, and can be expressed as (Perry et al., 1984):

$$Q_1 = c_o A \rho_1 [2((p-p_a)/\rho_1) + 2gH]^{1/2} \tag{4-4}$$

where: $Q_1$ = liquid mass emission rate (kg s$^{-1}$)

$c_o$ = coefficient of discharge (dimensionless)

$A$ = puncture area (m$^2$)

$\rho_1$ = liquid density (kg m$^{-3}$)

p = tank pressure (newtons m$^{-2}$)

$p_a$ = ambient pressure (newtons m$^{-2}$)

g = acceleration due to gravity (9.8m s$^{-2}$)

H = height of liquid above the puncture (m)


Critical to the analysis of liquid pipeline failures are the valve spacing and closing time, which are coupled to determine the total volume of the spill.  Long distances between the valves, or slow closing times, will allow a considerably greater volume of material to be released.  A pipeline will empty its contents from the failure point until the first closed valve.  Likewise, a storage tank failure will release liquid until the liquid level falls below the puncture level; a gas phase release will then occur as described in Section 4.1.1.


### 4.1.3  Two Phase Jet Releases


It is becoming more and more apparent that most emissions from pressurized tanks are two phase jets, consisting of a mixture of gas and liquid.  Thus the emission rate is somewhere in between that for a gas and a liquid, and can be calculated using empirical correlations recently developed by, for example, Fauske (1985) and Leung (1986).  These methods are only preliminary, however, and have not yet been incorporated into more comprehensive models.  Details on the AIChE Design Institute for Emergency Relief Systems (DIERS) study of emergency relief vent sizing are given by Swift (1984).  Leung (1986) suggests the following empirical equation for an isentropic flashing process:


For $\omega \geq 4$,


$$Q/(A(p_o\rho_o)^{1/2}) = (0.6055 + 0.1356 \ln \omega - 0.0131 (\ln \omega)^2)/\omega^{1/2} \qquad (4\text{-}5)$$


For $\omega < 4$,


$$Q/(A(p_o\rho_o)^{1/2}) = 0.66 \omega^{-0.39} \qquad (4\text{-}6)$$

where $\omega$ is an empirical "correlating parameter" developed from laboratory data and defined by:

$$\omega = x_0 \rho_0 / (\rho_1 - \rho_g) + p_0 \rho_0 A^2 / Q_1^2 \qquad (4-7)$$

where $x_0$ is the initial vapor mass fraction and subscript o refers to the condition in the tank. The parameter $(\rho_1 - \rho_g)$ is the difference between the liquid and gas densities for the chemical.

Fauske (1985) uses the following empirical equation for two phase choked flow releases from pressurized ammonia tanks or pipes:

$$Q = 0.943 \ (AL/(\rho_g^{-1} - \rho_1^{-1}))(T_b c_1)^{-0.5} \qquad (4-8)$$

where $T_b$ = Normal boiling point ($^O$K)
$\quad\quad\quad c_1$ = Liquid heat capacity (j kg$^{-1}$ $^O$K$^{-1}$), and
$\quad\quad\quad\ L$ = Heat of vaporization of the liquid (j kg$^{-1}$)

Researchers find that a pipe length on the order of 10 cm or greater is required between the source vessel and the emission point for the two phase flow to be developed to an equilibrium stage. It is stressed that these equations have not yet been tested against a wide range of chemicals and conditions. More importantly, there are no correlations currently available to estimate air entrainment into a two-phase momentum jet.

### 4.1.4 The Flashing Process

A significant fraction of a liquid spill may flash, depending on the normal boiling point of the liquid. Flashing is usually significant for any liquids, such as methane or ethane, that have normal boiling points well below ambient temperatures. One simple approach to determine the amount flashed is to assume that the vaporization process occurs so fast that it can be considered adiabatic. Thus, the heat of vaporization is provided solely from the heat capacity of the released liquid. For a single component and a constant temperature this heat balance yields a simple expression for the fraction of liquid flashed:

$$Q_f/Q_l = c_p(T-T_b)/L \qquad\qquad (4\text{-}9)$$

where $\quad Q_f$ = mass emission rate of liquid that flashes (kg s$^{-1}$)

$Q_l$ = total liquid mass emission rate (kg s$^{-1}$)

$c_p$ = heat capacity of the liquid (averaged between T and $T_b$) (j kg$^{-1}$ $^{\circ}$K$^{-1}$)

$T$ = temperature of the liquid in the tank ($^{\circ}$K)

$T_b$ = normal boiling point of liquid (assumed to be lower than T) ($^{\circ}$K)

$L$ = heat of vaporization of the liquid (j kg$^{-1}$)

For binary or multicomponent mixtures the flashing process is more complex and involves iterative flash calculations and knowledge of the vapor-liquid equilibrium behavior of the mixture.

In the case of a pressurized liquid pipeline break, some of the fluid, (liquid at the initial operating pressure), rapidly vaporizes or flashes to its gas phase as the pressure drops. As the pressure is relieved upstream, more liquid flashes to vapor. Figure 4-2, reproduced from Drivas et al. (1983), shows this occurring pictorially. At time $t_1$, the pressure wave has moved only a short distance, and most of the upstream fluid does not realize that a break has occurred. By time $t_2$, the pressure wave has traversed farther upstream and more flashing occurs.

For both storage tank and pipeline liquid releases, as the liquid and flashed gases flow out of the puncture into the atmosphere, they will entrain ambient air, which both dilutes and warms the cloud. Because of high velocities and the vigorous flashing process, the liquid may exit to a large extent in droplet form. This process may result in considerably more material being transported downwind as would be calculated by adiabatic flash calculations. Present models cannot adequately quantify this process. After the cloud becomes warmed by entrained air, more of the droplets will vaporize, again cooling the cloud. This cooling process, forming a dense cloud, should be accounted for in the atmospheric dispersion model.

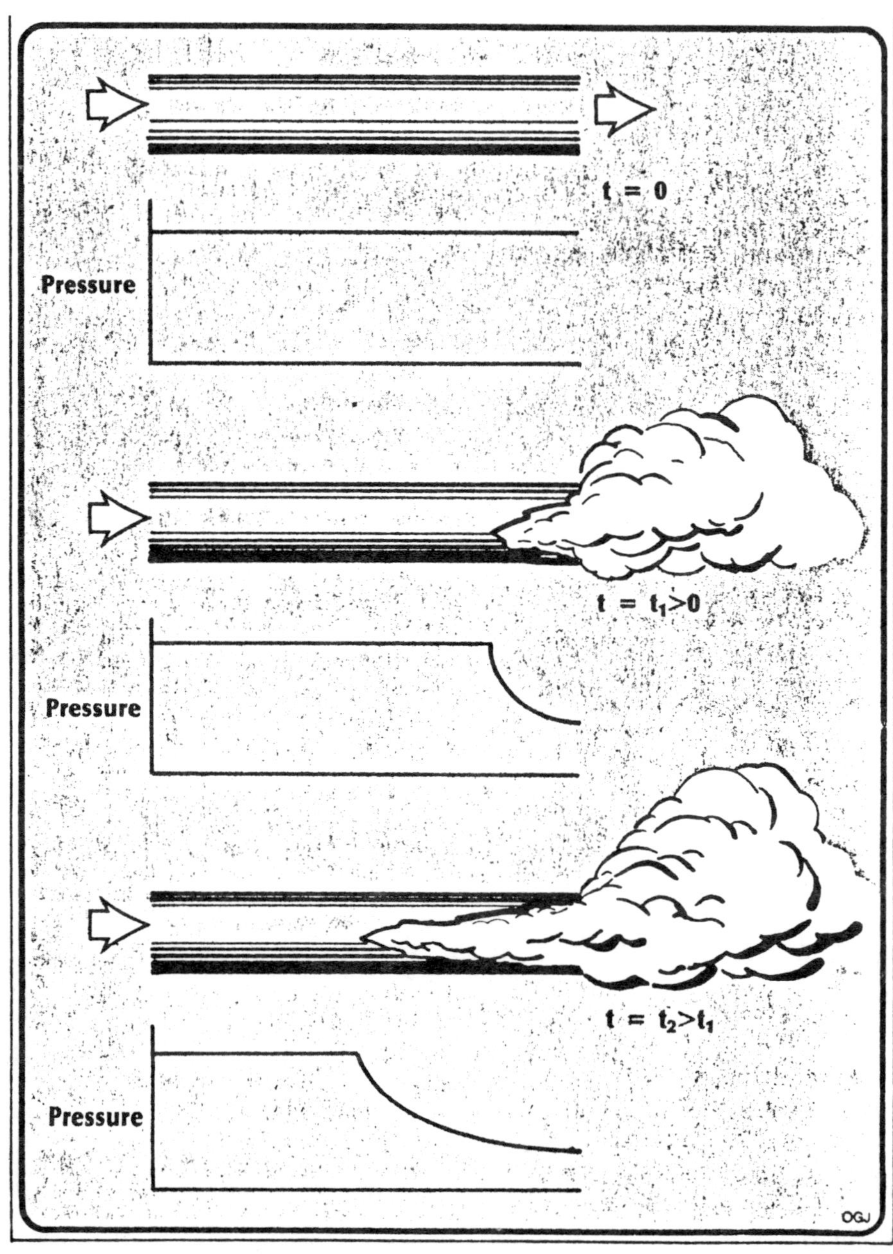



Figure 4-2.  Pressure histories after liquid pipeline break, showing flashing progressing upstream, from Drivas et al. (1983).

26

4.1.5  Liquid Pool Evaporation (Single Component)

The cases of gas, liquid and two-phase jet releases that have been described in the above three sections can be classified as time-varying releases, where the release rate is typically a decreasing function with time. These types of failures result from punctures in storage tanks or pipelines. Also of importance is the case of a failure of a vessel tank which results in the formation of a liquid pool. Because of the obvious combustion danger of large liquefied natural gas (LNG) spills, a considerable amount of effort has been spent on modeling the spreading and evaporation of cryogenic liquid pools. The subject of gaseous emissions due to evaporation is also important for hazardous waste dumps and landfills, where the emission rate may be low but the accumulated dosage over a long period of time may be significant.

In the case of liquid spills where the normal boiling point of the liquid is above ambient, a liquid pool will form on the ground and will slowly evaporate at a rate depending on its vapor pressure, substrate type, and atmospheric conditions. When a major cryogenic accident occurs, where the normal boiling point of the liquid is below ambient, the fluid may pour out quickly on the ground forming a large and very cold liquid pool. As the pool spreads across the ground, it absorbs heat and boils, producing a dense vapor cloud. When the pool is boiling, the evaporation is essentially limited by heat transfer into the pool, i.e. all heat input results in more vaporization.

The primary source of heat in the initial stage of evaporation is conduction through the ground. Convection from the air, incident solar energy, and radiative heat transfer can also add to the heat input of an evaporating pool. Figure 4-3, reproduced from Shaw and Briscoe (1978), shows the general heat balance of a liquid spill.

The conduction heat transfer from the ground is usually calculated by solving the standard one-dimensional heat flow equation with appropriate boundary conditions. Because the heat of vaporization is provided primarily by conduction, the ground below the boiling cryogenic liquid normally becomes colder, and ice formation in wet soil is a possibility. As an example of this extreme ground cooling, Figure 4-4, reproduced from Koopman et al. (1984), presents experimental data on temperatures taken very close to ground level

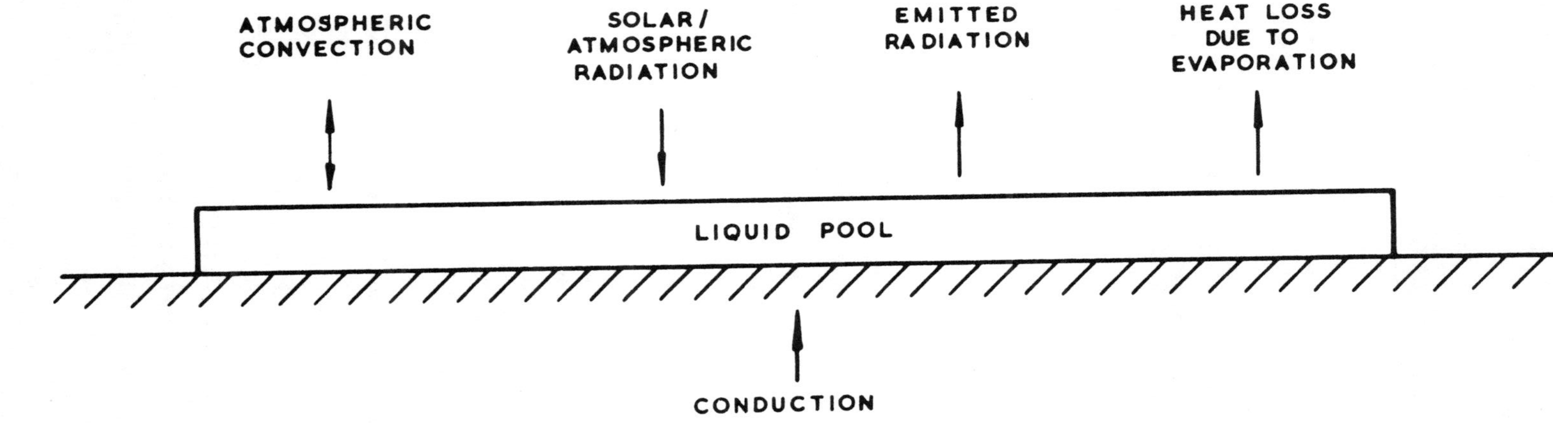


Figure 4-3.  The Heat Budget of an Evaporating Pool, from Shaw and Briscoe (1978).

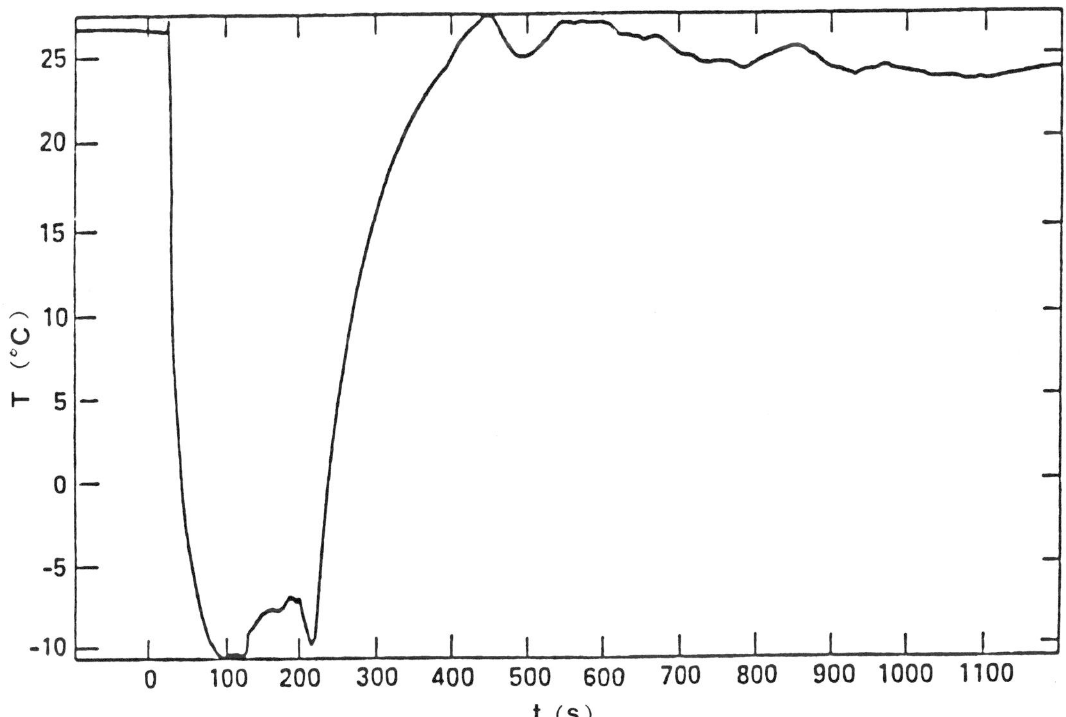

Figure 4-4.  Eagle $N_2O_4$ spill area temperature data at 2 cm above
ground level.  From Koopman et al (1984).

during a controlled nitrogen tetroxide ($N_2O_4$) spill.  The cold ground temperatures and especially ice formation in wet soil can significantly decrease heat transfer to the pool, and thus decrease the evaporation rate.

When the temperature of a liquid pool is lower than its normal boiling point, some models assume that the evaporation is limited by mass transfer from the liquid pool to the gas phase.  The evaporation rate for a single component liquid, assuming an ideal gas phase, either a shallow pool or a well-mixed deep pool, and a steady state condition, can be expressed as (Fleischer 1980):

$$Q_a = k_g A_p p_s m / (R^* T_a) \qquad\qquad (4\text{-}10)$$

where   $Q_a$ = gas mass emission rate   (kg $s^{-1}$)

$k_g$ = mass transfer coefficient (m $s^{-1}$)

$A_p$ = pool area ($m^2$)

$p_s$ = vapor pressure of the compound at temperature $T_a$ (newtons $m^{-2}$)

$m$ = molecular weight,

$R^*$ = gas constant (8.31436 joules mole$^{-1}$ $^{O}K^{-1}$)

$T_a$ = ambient temperature ($^{O}K$)

This equation is conservative for rapidly evaporating liquids, since it assumes no material is present in the bulk air above the liquid.  To calculate the mass transfer coefficient, $k_g$, in equation (4-10), there are two main methods that are used in source emission models.  One is a chemical engineering approach, which uses the general relationship (Perry et al., 1984):

$$N_{Sh} \equiv k_g d / D_m = f(N_{Re}, N_{Sc}) \qquad\qquad (4\text{-}11)$$

where   $N_{Sh}$ = Sherwood number

$d$ = effective pool diameter (m)

$D_m$ = molecular diffusivity of the compound (units $m^2$ $s^{-1}$),

$N_{Re}$ = Reynolds number, and

$N_{Sc}$ = Schmidt number (= kinematic viscosity, $\upsilon$, divided by the molecular diffusivity, $D_m$).

Typically, standard chemical engineering empirical correlations, based on laminar or turbulent flow over flat plates, are used to calculate $k_g$ in equation (4-11), and Fleischer (1980) gives several examples (see eq. 4-21) in his discussion of the SPILLS model.

The other main method to calculate the mass transfer coefficient, $k_g$, for slowly evaporating pools is a meteorological approach based primarily on the work of Sutton (1953), who solved the steady-state atmospheric diffusion equation over a liquid pool with a power-law velocity profile and a corresponding power-law eddy diffusivity profile. An expression for $k_g$ as a function of windspeed, atmospheric stability, and liquid pool dimension has been used by Mackay and Matsugu (1973), who formed a correlation based on experimental data on evaporation for neutral atmospheric stability:

$$k_g = 0.00482 \ N_{Sc}^{-0.67} u^{0.78} d^{-0.11} \qquad (4\text{-}12)$$

where u is windspeed in $ms^{-1}$, d is depth of pool (m), and $k_g$ is in $m \ s^{-1}$. This expression has been commonly used in source emission models to calculate $k_g$ for the evaporation rate from single component pools.

If the liquid boiling point is above the ambient temperature, then standard formulas for evaporation fluxes can also be used, such as the mass transfer equation (Hanna et al., 1982):

$$Q_a = C_D u \ (\rho_{gs} - \rho_g) \qquad (4\text{-}13)$$

where $\rho_{gs}$ is the saturation density of the gas at ambient conditions and $\rho_g$ is the actual density of the gas at a height of 10m above the pool. The drag coefficient, $C_D$, is about $10^{-3}$ if u and $\rho_g$ are measured at a height of 10m.

Liquid spills on water will generally increase in area with time, as will unconfined spills on land, caused primarily by gravity spreading. Shaw and Briscoe (1978) present a good summary of the equations used to calculate how unconfined spills increase in area on both land and water. However, storage tanks on land normally have some form of confinement around them in the event of an accident.  This may be a dike, a dirt embankment around the tank, or in some cases, the area surrounding the tank is dug out to hold the tank contents in the event of a failure.

This confinement can be extremely important in limiting the emission rate from a storage tank failure.  As shown by equation (4-10), total evaporative emission rates are directly proportional to the area of the spill, and thus limiting the pool area is a very effective method to limit source emission rate.

Figure 4-5, reproduced from Shaw and Briscoe (1978), presents some calculations of source emissions as a function of time for both confined and unconfined cases of a 1000 m$^3$ LNG spill on land.  For both instantaneous and continuous (finite rate) spills, the confined source emissions at a given time are roughly a factor of five less than the unconfined emissions.

### 4.1.6  Liquid Pool Evaporation (Multi-Component)

Many potentially-hazardous liquids are multicomponent, and hence cannot be treated by the single equations given in the last subsection.  Drivas (1982) has used these equations for single components to derive equations for multicomponents.  Assume that $Q_a$ is the total evaporative emission rate (kg m$^{-2}$ s$^{-1}$) of all components.  Then the following empirical equation is derived:

$$Q_a = (k_g M_T{}^o / (n_T R^* T)) \sum_{i=1}^{N} (x_i{}^o p_{is} m_i e^{-kp_{is}t}) / \sum_{i=1}^{N} x_i{}^o m_i \qquad (4\text{-}14)$$

where $k_g$ is the mass transfer coefficient for the gas phase defined by equation (4-12),

$M_T{}^o$ is the total initial mass of evaporable liquid (kg)

$n_T$ is the total moles of liquid

$x_i{}^o$ is the initial liquid mole fraction of component i

$p_{is}$ is the saturation vapor pressure of component i (newtons m$^{-2}$)

$m_i$ is the molecular weight of component i (kg mole$^{-1}$)

$k = k_g A_p / (n_T R^* T)$

$R^*$ is the universal gas constant (8.31436 joules mole$^{-1}$ $^o$k$^{-1}$)

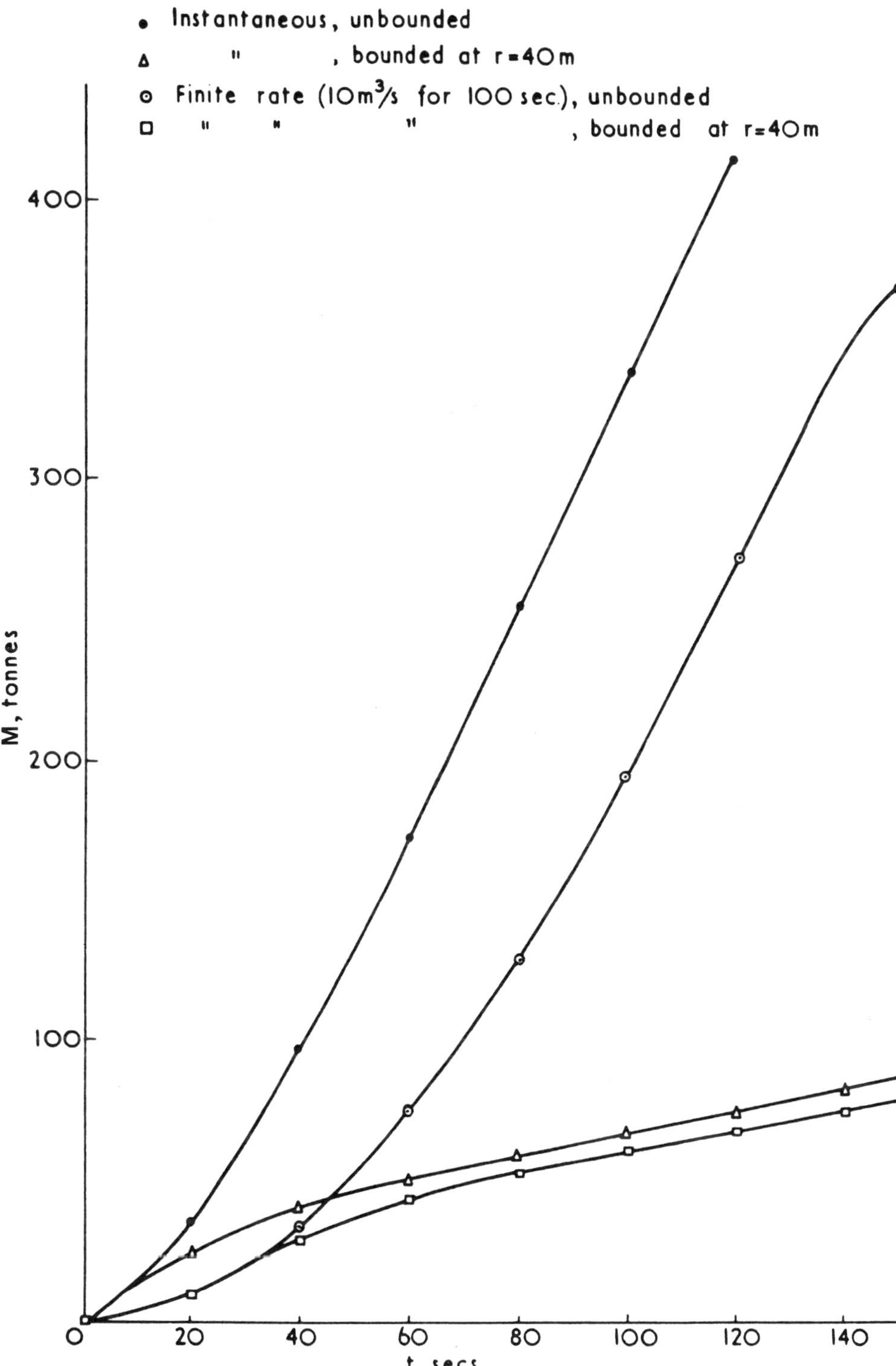


Figure 4-5. Calculation of mass vaporized M at time t for a spill of 1,000m³ (416 tonnes) LNG on average soil, from Shaw and Briscoe (1978). A large difference between bounded and unbounded spills is evident.

The initial fraction $x_i^o$ can be derived from a knowledge of the initial liquid
composition or from observations over the liquid, using Raoult's law.  It is
assumed that mass transfer in the liquid is sufficiently rapid that total
transfer is dominated by the gas phase (equation 4-14).  Drivas (1982) verifies
this equation by comparisons with observations of evaporation from a
multicomponent oil spill.  However, there is clearly much more work required to
refine these procedures and incorporate them into applied models.

The only practical model that currently handles binary evaporation is that
of Wu and Schroy (1979).  Their model is equivalent to equation (4-14) for N
equal to 2, and they account for resistance to mass transfer in the liquid
phase.  They have validated their model using literature data for mixture
evaporation.

4.2  Review of Procedures Used by Specific Models for Predicting Source
     Emission Rates

This section presents a summary of several currently available practical
or applied models for calculating source emission rates and provides a general
classification of models by their capability for calculating important types of
emission release situations.  For each type of source release category, one
specific model, which contains algorithms that are typical of the models in
that class, will be described in detail.  Finally, a brief summary will be
presented of previous studies that compare calculated source emission rates
from different models.

A list of some available source emission models is presented in Table 4-1,
which segregates model capabilities into jet releases and liquid pool
evaporation.  This list contains only models that have been reported in the
open literature.  Appendix A contains a more detailed summary of the
capabilities of the models for which authors completed a questionnaire,
including some proprietary models.  None of the models treats two-phase jets,
and only the Wu and Schroy (1979) model treats binary-component evaporation.

Some models can predict more than one type of emissions; these models have
been developed to handle a variety of emission sources, but are usually more
complicated and more difficult to use than models that focus on a single type

of source.  Most of the models that can predict pool spreading are applicable only to cryogenic releases.

### 4.2.1  Gas Jet Release Models

Five of the models listed in Table 4-1 have the capability to predict gas jet releases.  These are the pipeline model of Wilson (1979), the general model developed by the Ontario Ministry of the Environment (MOE 1983), the CHARM model (Eltgroth et al. 1983), the AIRTOX model (Paine et al. 1986), and the DENZ model (Fryer and Kaiser 1979).  Note that these are single-phase models.

All five gas release models are based on the standard choked or critical flow equation, as presented in equation (4-2), with various techniques to calculate the changing pressures and temperatures to use in that equation.  The pipeline model of Wilson (1979) presents detailed equations for determining the pressure and temperature to use in equation (4-2), for the assumption of either isothermal flow or adiabatic flow.  For either case, numerical iterative calculations are necessary to correctly predict the pressure and temperature.  The Wilson (1979) model is discussed here as an example of the physical assumption in gas jet models.

Wilson (1979) also presents a simpler empirical model developed by Bell (1978) that compares quite well with the more detailed numerical calculations.  This empirical model expresses a pipeline gas release as a "double exponential" which decreases with time with two important time constants:

$$Q = (Q_0/(1+(M_0/\beta Q_0)))(\exp(-t\beta Q^2_0/M_0^2)+(M_0/\beta Q_0)\exp(-t/\beta)) \qquad (4-15)$$

where  $Q$ = gas mass emission rate (kg s$^{-1}$)
$\quad Q_0$ = initial gas mass emission rate, calculated from equation (4-2)
$\qquad$ using the initial gas temperature and pressure,
$\quad M_0$ = initial total mass of gas in the pipeline (kg), and
$\quad t$ = time(s).

The parameter, $\beta$, is a time constant given by:

$$\beta = 0.67 \, (\gamma f L_p/d_p)^{1/2} L_p/u_s \qquad (4-16)$$

Table 4-1.  List of Some Available Source Emission Models
More Details are Given in Appendix A


Evaporation Models:

Ille and Springer (1978)

Army (Whitacre et al. 1986)

Shell SPILLS (Fleischer 1980)

USAF ESL (Clewell 1983)

Air Weather Service (AWS 1978)

Illinois EPA (Kelty 1984)

Stiver and Mackay (1982)

Monsanto (Wu and Schroy (1979)

Shaw and Briscoe (1978)


Jet Model:

Wilson (1979)


Jet and Evaporation Models:

CHARM (Eltgroth et al. 1983)

Ontario MOE (MOE 1983)

AIRTOX (Paine et al. 1986)

Kunkel (1983, 1985)

DENZ (Fryer and Kaiser 1979)

COBRA (Alp 1985)

where  $\gamma$ = gas specific heat ratio

   $f$ = pipe friction factor

   $L_p$ = length of pipe from break to first closed valve (m)

   $d_p$ = pipe diameter (m), and

   $u_s$ = sonic velocity in the gas = $(\gamma R^* T/m)^{1/2}$   $(ms^{-1})$

## 4.2.2 Liquid Jet Release Models

There are a number of models in Table 4-1 that have the capability to predict liquid releases and others are listed in Appendix A.  These include the CHARM model (Eltgroth et al., 1983), the AIRTOX model (Paine et al. 1986), the Ontario MOE model (MOE 1983), and the DENZ model (Fryer and Kaiser 1979).

In general, the models use either Torricelli's principle (for example, equation (4-4)) or a specific solution of Bernoulli's equation to calculate the rate of liquid release from a small puncture in a liquid storage tank.  As a typical example, the Monsanto model of Wu and Schroy (1979) calculates the rate of liquid spill by using equation (4-4) and the following expression for the liquid height, H, for vertical tanks:

$$H = 4V_1/(\pi d_t^2) \tag{4-17}$$

where $V_1$ = liquid volume remaining in the tank, and

   $d_t$ = tank diameter.

For horizontal tanks, H is calculated in the Monsanto model on a dynamic basis using the following equations:

$$H = d_t (1 - \cos\theta)/2 \tag{4-18}$$

$$V_1 = .25\, L_t d_t^2 [\theta - \sin(2\theta)/2] \tag{4-19}$$

where  $\theta$ = the angle relative to the upward vertical direction formed by the liquid surface remaining around the circumference of the tank (always greater than $90^{\circ}$), and

   $L_t$ = length of the horizontal tank (m)

### 4.2.3  Evaporation Models - Constant Pool Area (Single Component)

A large number of source emission models have been developed for the important case of liquid spill evaporation assuming a constant liquid pool area.  The models exhibit a wide variety of algorithms for calculating evaporation, but in general use some form of the basic   relationship presented in equation (4-10), with the emission rate proportional to pool size and the vapor pressure of the compound.  The models in the table are for liquids with single components.  Generally, infiltration into the soil surface is neglected for cryogenic liquids because of the frost layer rapidly established beneath the surface.

As discussed in Section 4.1.5, the techniques used to calculate the mass transfer coefficient, $k_g$, in equation (4-10) are based primarily on either the chemical engineering approach in equation (4-11) or a meteorological approach such as equation (4-12).  Some of the simpler models use a completely empirical expression.  Calculated source emissions from a number of these evaporation models are compared in Section 4.2.5.

An example of a sophisticated and widely used evaporation model for single components is the Shell SPILLS model (Fleischer 1980).  This model has the important capability of heat transfer calculations for steady-state evaporation from spills.  Many persons have adapted this model to allow use for chemicals other than those originally intended.  To calculate the conduction heat flow rate into a boiling pool from the ground, a solution of the standard one-dimensional heat transfer equation is used, resulting in:

$$H_{ch} = k_s A_p (T_o - T_b) / (\pi \alpha_s t)^{1/2} \tag{4-20}$$

where $H_{ch}$ = conduction heat flow into the pool (watts)

$\quad\quad k_s$  = thermal conductivity of the soil (watts $m^{-1}$ $^o K^{-1}$)

$\quad\quad A_p$  = pool area ($m^2$)

$\quad\quad T_o$  = initial soil temperature ($^o K$)

$\quad\quad T_b$  = boiling point of the liquid ($^o K$)

$\quad\quad \alpha_s$  = thermal diffusivity of the soil ($m^2$ $s^{-1}$), and

$\quad\quad t$  = time (s)

Corrections can be made to $k_s$ and $\alpha_s$ for the case of ice formation in the soil.

When the temperature of the liquid is below its boiling point, the Shell SPILLS model assumes evaporation limited by mass transfer using equation (4-10). The mass transfer coefficient is calculated by using standard chemical engineering correlations for laminar and turbulent flow over a flat plate, assuming that the laminar to turbulent flow transition occurs at a Reynolds number of 320,000 based on pool diameter:

$$k_g d/D_m = N_{Sh} = 0.037\, N_{Sc}^{1/3}[N_{Re}^{0.8}-15,200] \tag{4-21}$$

where: $k_g$ = mass transfer coefficient $(ms^{-1})$
   d = effective pool diameter (m)
   $D_m$ = molecular diffusivity of the compound $(m^2 s^{-1})$
   $N_{Sh}$ = Sherwood number
   $N_{Sc}$ = Schmidt number, and
   $N_{Re}$ = Reynolds number

4.2.4  Evaporation Models - Spreading Pool Area (Single Component)

Table 4-1 lists some practical models that can calculate evaporation rates from single-component liquid pools that are increasing in area with time. These are the Monsanto model (Wu and Schroy 1979), the Shaw and Briscoe (1979) model, the DENZ model (Fryer and Kaiser 1979), and the COBRA model (Alp 1985). With the exception of the Monsanto model, these models have been developed specifically for cryogenic spills and consider only evaporation limited by heat transfer. These models then are applicable primarily to spills of compounds with boiling points that are well below ambient.

As an example of this type of model, Shaw and Briscoe (1978) have developed a spreading pool model for the Safety and Reliability Directorate of the United Kingdom Atomic Energy Authority. The model is selected for discussion because of the complete and readable discussion in the reference. This model was derived for cryogenic spills and assumes that, for spills on land, heat transfer into the boiling liquid pool is limited by conduction from the soil. A set of four simultaneous heat and mass conservation equations is solved analytically by Shaw and Briscoe (1978) for the gravitational phase of spreading to determine the pool area increase with time for an instantaneous spill on land:

$$r = ((8gV_o/\pi)^{1/2}t + r_o^2)^{1/2} \qquad (4\text{-}22)$$

where  $r$ = pool radius, (m)

$g$ = acceleration due to gravity, $(9.8 \text{ ms}^{-2})$

$V_o$ = initial volume of spill, $(m^3)$

$t$ = time from spill (s), and

$r_o$ = initial radius of contained liquid (m)

Usually $V_o$ and $r_o$ are approximations based on empirical evidence.

The corresponding mass vaporized, M, at time t is:

$$M = 8S_r k_s (T_o - T_b)(2\pi gV_o)^{1/2}t^{3/2}/(L(\pi\alpha_s)^{1/2}) \qquad (4\text{-}23)$$

where  $S_r$ = soil "roughness" factor (to correct for ice formation, for example).  This factor is of order one.

and    $L$ = heat of vaporization of the liquid (joules $kg^{-1}$)

Shaw and Briscoe (1978) also present similar derivations for continuous cryogenic spills on land and for instantaneous and continuous spills on water. These formulas are for flat terrain.  Evaporation is likely to be larger for sloping terrain, since the initial pool will flow downhill to cover a broader area.

4.2.5  Comparison of Source Model Calculations

Two good reviews have been published that describe and compare a number of available source emission models -- Kunkel (1983) and McNaughton et al (1986). Kunkel (1983) has reviewed and used five different models -- Ille-Springer (I&S), Army, Shell, Air Force Engineering and Services Laboratory  (ESL), and Air Weather Service (AWS) -- for calculating evaporative emissions over a constant pool area for a single component. Calculations were made using each model (with the exception of the Shell SPILLS model) to predict the evaporative emission rate for hypothetical spills of four different chemicals.

The results are summarized in Figures 4-6 and 4-7, which are reproduced from Kunkel (1983). Each figure shows a separate graph of calculated emission rates for each of the four chemicals modeled ($N_2H_4$ - hydrazine; MMH - monomethylhydrazine; UDMH - unsymmetrical dimethylhydrazine; $N_2O_4$ - nitrogen tetroxide). In these two figures, the various model predictions are compared against each other and no comparisons are made with observed data.

Figure 4-6 shows the model calculations as a function of ambient air temperature. In general, the calculated emission rates of the various models increase similarly with temperature, because most of the models use an algorithm similar to equation (4-10), where the emission rate is directly proportional to the vapor pressure. The difference among the emission rates calculated by the models is approximately a factor of two, for ambient air temperatures above $10^{\circ}$ C (and discounting the Ille and Springer $70^{\circ}$ curve). Note that the AWS predictions are independent of ambient air temperature, but pass roughly through the middle of the other predictions.

Figure 4-7 shows the same model calculations as a function of windspeed. Because most of the models use a power-law windspeed relationship similar to equation (4-12), their calculated emission rates increase similarly with windspeed. Again, the difference among the emission rates calculated by the models is approximately a factor of two. In this case, the AWS predictions are an increasing function of wind speed.

McNaughton et al (1986) also reviewed a number of emission and dispersion models and presented some results of source model comparisons with four sets of experimental data. The calculated model emission rates are compared with the observed experimental data in Table 4-2, which is reproduced from McNaughton et al (1986). The two tests labeled "Edgewood" refer to small-scale experimental tests with nitrogen tetroxide ($N_2O_4$), and the two tests labeled "LPG-" refer to experimental liquefied propane spills.

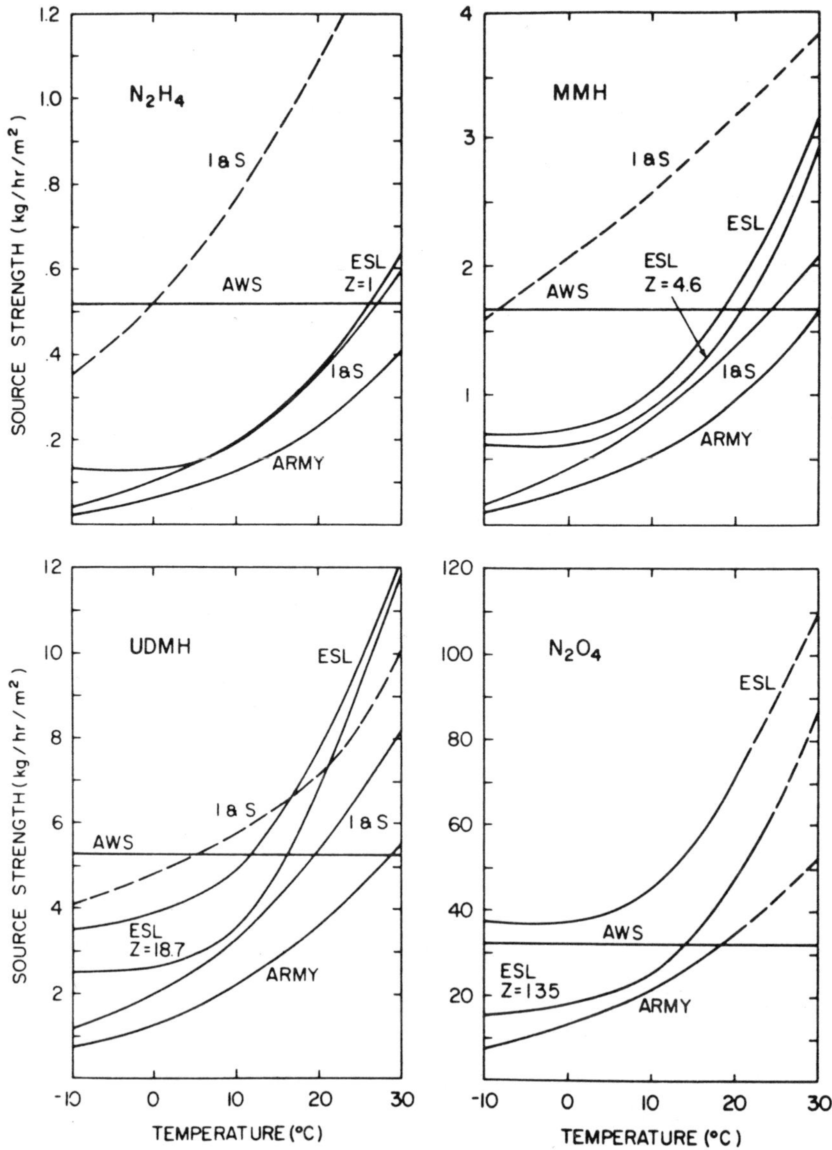



Figure 4-6.  Evaporative source strengths (kg $hr^{-1}m^{-2}$) as a function of ambient air temperature as determined from different models (from Kunkel 1983).  Wind speed is 2 m/sec.  Dashed line represents Ille and Springer model with 70° solar angle.

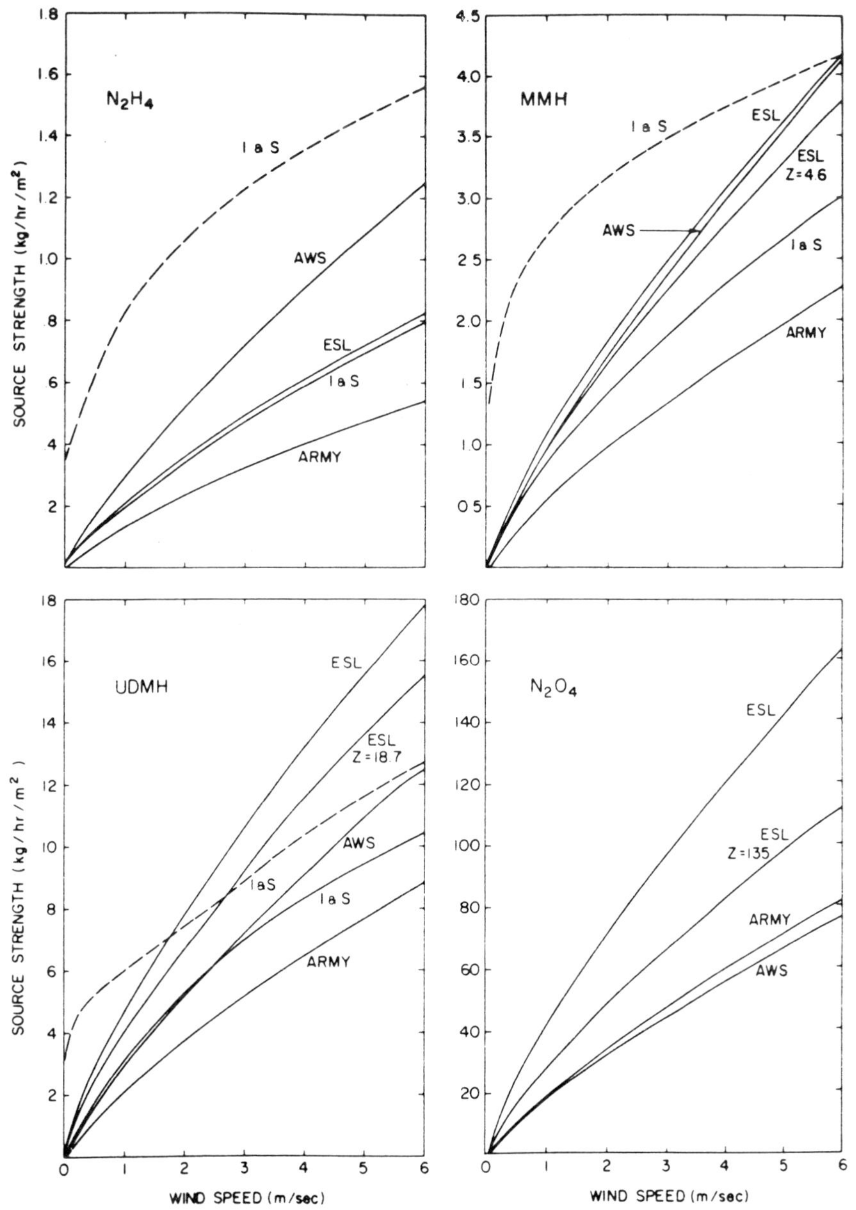



Figure 4-7. Evaporative source strengths (kg hr$^{-1}$m$^{-2}$) as a function of wind speed as determined from different models (from Kunkel 1983). Temperature is 20°C. Dashed line represents Ille and Springer model with 70° solar angle.

43

As shown in Table 4-2, the difference among the emission rates calculated
by the six models for the "Edgewood" $N_2O_4$ tests is approximately a factor of
three for both tests.  However, almost all the models greatly overpredicted
the observed emission rates for $N_2O_4$, with a median overprediction of a factor
of about three.

The results for the "LPG-" liquefied propane tests, although limited to
only two models, demonstrate considerable sensitivity to model input
assumptions.  For the Illinois EPA model, changing the pool temperature from
$-43^{\circ}C$ to $-51^{\circ}C$ resulted in an emission rate decrease of about a factor of
three.  For the Shell SPILLS model, the assumption of wet soil instead of dry
soil resulted in an emission rate decrease of a factor of seven in one case and
a factor of three in the other comparison.  These results demonstrate that a
proper determination of input parameters is extremely important.

Table 4-2

Summary of Predictions of Evaporation Rates (kg/hr) by Six
Models Using Data from Four Field Tests (from McNaughton et al 1986)

| Model | Test | | | |
| | Edgewood 6 | Edgewood 10 | LPG-CS37 | LPG-PL38 |
|---|---|---|---|---|
| Illinois EPA | 64 | 64 | 41.7[*] <br> 16.0[**] | 41.7 <br> 16.0 |
| AWS | 103 | 53 | n.a. | n.a. |
| Army | 172 | 88 | n.a. | n.a. |
| U.S.A.F. ESL | 59 | 31 | n.a. | n.a. |
| Ille & Springer | 49 | 24 | n.a. | n.a. |
| Shell Spills | 59 | 20 | 149 (21.5)[+] | 190 (57)[+] |
| Observation | 18.6 | 16.8 | 16.7 | 6.9 |

[*] Vapor pressure at $-43^\circ$C
[**] Vapor pressure at $-51^\circ$C
[+] Dry soil value; wet soil value in parentheses

# 5

# Transport and Dispersion Models

The source emissions models reviewed in Section 4 are mostly based on empirical engineering approximations to basic physical principles. Only a small fraction of the source models have been adequately evaluated with field data. The situation with transport and dispersion models for hazardous materials is similar, but the number of available models is much larger - greater than 100 and growing at a rate of about 10 per year. A few of these models are the result of long-term basic research programs in which an attempt is made to thoroughly understand the process. Many of these models are older inert-gas models which have been slightly retooled and now applied to hazardous materials. Other models have been constructed by patching together several independent algorithms. This situation must be very confusing to the person encountering the plethora of models for the first time, and is one of the reasons for preparing this book.

The response by authors to a request for information on these models has been overwhelming, and it is impractical to thoroughly cover each model. Instead, the models will be grouped into classes (e.g. dense gas box models) and the general physical principles applying to each class will be reviewed. A few specific models will then be discussed so that the reader can appreciate how the various assumptions are put together.

Nearly all of these models are in the process of being improved at any given time. Whether it is terrain effects, aerosols, entrainment, or dense gas slumping, there is _something_ that is being changed. If the statement were made that model A neglects process B, by the time this book reaches print, model A might well be changed. It is hoped that readers recognize this characteristic of a rapidly evolving research field.

Section 4 covered the source emissions region of the generalized hazardous gas modeling diagram that was previously drawn in Figure 2-1. This section will cover all aspects of transport and dispersion of the released material once it is in the atmosphere. As illustrated in Figure 2-1, four regions can be discussed:

Initial acceleration and dilution.

Dominance of internal buoyancy.

Transition from dominance of internal buoyancy to dominance of
ambient turbulence.

Dominance of ambient turbulence.

The initial acceleration phase has received very little attention in the
published literature and can be covered in this single paragraph. It is
important mainly for pressurized materials which are subjected to failure of
the container or pipeline, and for accidents in which explosions occur.
Kaiser and Walker (1978) provide several examples, such as the crash of a
tanker truck containing pressurized chlorine or ammonia. On the basis of
photographs and other empirical evidence, they suggest that the rapid expansion
or explosion results in an initial dilution in the range from 10:1 to 100:1.
It is obviously very difficult to accurately measure this parameter in a field
experiment. Even relatively well-behaved continuous momentum jets are
characterized by an initial "zone of flow establishment", as defined by Ooms
(1972).

The remainder of this section will first discuss the basic physical
principles of transport and dispersion models, and then review some procedures
used by specific models for transport and dispersion.

5.1  Discussion of Physical Principles

The simple diagram in Figure 2-1 represents all the important regions of
hazardous gas modeling, but does not do justice to the myriad of source and
sampling conditions. If the duration of the release is much less than the
transport time to a receptor, then the release can be assumed to be
instantaneous, since the recepter "sees" a more-or-less circular puff. If the
reverse is true, then the source emissions can be considered to be continuous.
Most accidental sources will be somewhere in-between, with a time-variable
release rate. Wilson (1981) has obtained an interesting analytical solution
for a release rate that decreases exponentially with time. The so-called

instantaneous puff models can simulate a variable release rate by assuming discrete increments in emissions.

Similarly, little effort has gone into suitable definitions of sampling conditions. The averaging time and the averaging volume are important, since peak concentrations tend to increase as the averaging time or volume decreases. Most models give no indication of the intended averaging periods for their calculations. A model without inclusion of this information makes interpretation of model results difficult, since health effects for hazardous materials cover averaging periods from one second to several hours. In some cases total dosage (concentration integrated over time) is important. The last subsection below will deal with the effects of averaging times and volumes and the related subject of concentration fluctuations.

The following paragraphs will provide a comprehensive review of the basic physical principles involved in hazardous material modeling in the atmosphere. Regions 3 through 5 in Figure 2-1 will be covered.

5.1.1  Initial Momentum Jets and Buoyant Plumes

It is clear from field observations and theoretical arguments that ground level concentrations of toxic pollutants can be substantially reduced if the initial height of the plume above the ground can be increased. This fact has led to the construction of 300 to 400m stacks at a cost of millions of dollars at power plants and other industrial facilities. With accidental releases, the emission height can be controlled only to the extent that elevated safety vents are constructed. The Gaussian plume equation for neutrally-buoyant gases can be used to quantify the benefits of increased plume height, $z_e$:

$$C = (Q/\pi u \sigma_y \sigma_z) \exp(-y^2/2\sigma_y^2) \exp(-z_e^2/2\sigma_z^2) \tag{5-1}$$

where the parameters are defined in the following way:

C  = Ground level concentration ($kg/m^3$)

Q  = Continuous source strength (kg/s)

u  = Wind speed (m/s)

$\sigma_y$ and $\sigma_z$ = Lateral and vertical dispersion parameters (m)

$y$ = Lateral distance from plume center (m)

$z_e$ = Height of plume center above ground (m)

Methods of estimating $\sigma_y$ and $\sigma_z$ are discussed in section 5.1.3.

If the emissions have a significant velocity or if the gas density is different from the atmospheric density, then the height $z_e$ can be different from the initial height of the release. Methods for calculating this plume rise component are given later for momentum jets and dense plumes, and are given by Hanna et al. (1982) for positively buoyant plumes. Equation (5-1) is valid at downwind distances ranging from near the source to 10km for industrial stacks and would be valid for hazardous gas releases that have neutral density and are continuous. For special conditions ($\sigma_y \propto \sigma_z$), the maximum occurs where $2\sigma_z^2 = z_e^2$ and has the value:

$$C_{max} = (\emptyset.74Q/\pi z_e^2 u)\,(\sigma_z/\sigma_y) \tag{5-2}$$

which clearly shows the dependence of the ground-level $C_{max}$ on the inverse square of the plume height $z_e$ (Hanna et al. 1982). If plume height doubles, then maximum concentration at the ground decreases by a factor of four. In the case of dense gas plumes, this benefit is partially cancelled by the tendency of the plume to sink towards the ground. Thus any means that can be used to increase the initial plume elevation can be very effective in reducing ground level concentrations.

## Releases that Can be Modeled with Standard EPA Procedures

Several decades of research have gone into the development of plume rise models for typical industrial sources. The EPA UNAMAP 6 air quality models (available from NTIS) contain algorithms for the following simultaneous conditions:

1)    Continuous releases

2)    Momentum jets

3)    Plumes with neutral or positive buoyancy

A continuous release is one for which the release rate is constant over a time period exceeding the averaging time at the receptor, or for which the duration of the release exceeds the travel time from source to receptor. These plume rise models are extensively reviewed in other documents (e.g. ch. 2 of Hanna, Briggs and Hosker 1982; Briggs 1984; Users Guides for EPA UNAMAP6 Models) and will not be reviewed here. Suffice it to say that two fundamental source parameters must be estimated:

$$F_o = g((T_p-T_a)/T_p)w_oR^2_o \qquad \text{Buoyancy Flux} \qquad (5-3)$$

$$M_o = w^2_oR^2_o \qquad \text{Momentum Flux} \qquad (5-4)$$

where $w_o$ and $R_o$ are initial plume velocity and stack radius, and $T_p$ and $T_a$ are initial plume and ambient temperatures (differences in molecular weights are neglected in equation 5-3). Note that these EPA models do not treat many problems of concern to hazardous material modelers, such as dense gases and non-continuous (i.e., instantaneous or transient) releases.

## Momentum Jets and Dense Gas Plumes

Two types of special plume rise models have been studied for continuous hazardous gas releases. The first type of model is for momentum jets from pipeline ruptures and the second type of model is for dense gas plumes from elevated releases. The momentum jets that result from many types of pressurized pipeline and storage tank ruptures have been most thoroughly studied by Wilson (1979, 1981), and his model has been subsequently used by Blewitt (1985) and Leahey and Schroeder (1986). Figure 5-1 presents drawings of several types of high-momentum releases, which lead to the conclusion that this type of modeling is highly uncertain because one seldom knows in which direction the jet will blow. The worst case may be the lateral plume that flows parallel to the ground, since it experiences little dilution but remains near the ground. The relative impact of these different scenarios also depends on the receptor location (direction and distance from the rupture).

Figure 5-1.  Types of momentum jet releases, illustrating the various directions the jet may blow and the deflections off the ground surface that may occur.

**5. Transport and Dispersion Models**

●   Momentum Jet


Wilson (1979, 1981) assumes that choked flow exists with sonic velocity at the pipe rupture (Section 4 covers the dynamics of the flow at the point that it passes through the source orifice, and this section covers the dynamics of the flow in the ambient atmosphere).  His theoretical analyses and the results of field experiments suggest that the expansion and recompression process has no effect on the jet temperature.  The "plume rise," or distance that the plume travels in a crosswind, $\Delta h$, is also assumed to be dependent on the current value of the release rate, since the time for the plume to travel from the rupture to a final $\Delta h$, is usually much less than the time scale of the decrease in release rate as the pipeline or tank empties.  He suggests the following equation for the plume rise or the distance $\Delta h$ travelled by the jet in a crosswind u:


$$\Delta h = 4.8 \ M_o^{1/2}/u \tag{5-5}$$


where $M_o$ is the initial momentum flux (divided by $\rho\pi$) in $m^4 s^{-2}$ out of the pipe, the wind speed u has units m/s and the distance $\Delta h$ has units m.  Note that there are no restrictions on whether the jet is pointed upward or downward; only that the jet must be perpendicular to the wind flow and there must be no obstructions to the jet.  Field experiments in Alberta with pipeline ruptures directed upward show that equation (5-5) is accurate within a factor of about two (Wilson, 1979) for jets pointed upward.  For the case of $H_2S$, $M_o$ (in $m^4 s^{-2}$) equals 89 times the mass emission rate (in $g \ s^{-1}$).


The jet is found to expand at a rate such that its radius equals about 0.4 times its distance from the source.  Thus the horizontal jet in Figure 5-1 will be inhibited by the ground and the effects of increased air entrainment may lead to equation (5-5) being an overestimate.  Similarly, equation (5-5) does not apply to the lower two examples in the figure, where the jet interacts with the ground and which can best be parameterized by physical modeling or full-scale field experiments. In the meantime, these two examples can perhaps be parameterized by assuming zero plume rise and an initial dilution of about a factor of ten. Beyond a distance of about 100m, the jet effects are rather unimportant for typical pipeline ruptures.

● Elevated Dense Gas Plumes

The second type of special plume rise model is intended to deal with continuous elevated releases of dense gases. Bodurtha (1961), Ooms et al. (1974) and Hoot, Meroney and Peterka (1973) have pioneered in the study of this phenomenon, and Ooms and Duijm (1984) provide a useful review of the subject. The EPA has recently been testing the RVD model, which is based on the Hoot et al. (1973) equations, but multiplies the predicted concentrations by an arbitrary factor of five. A good example of the problem that must be modeled is seen in Figure 5-2, which contains observations from a wind tunnel experiment by Xiao-Yun, Leijdens and Ooms (1986). The dense gas plume rises at first due to its initial upward momentum from the stack, but then sinks due to its excess density . Eventually the plume centerline strikes the ground surface.

The available models for dense gas plume trajectories are more complex than the plume rise algorithms in the EPA UNAMAP6 models, which are based on simple solutions derived by Briggs (1975, 1984). The Hoot et al. (1973) and Ooms et al. (1974) models are based on the equations of mass, momentum, and energy conservation, and require either integrations or piecewise analytical solutions. All models also require entrainment assumptions (regarding the rate at which the plume is diluted by ambient air) to close the system and thus permit a solution. These models are also restricted to gas-phase plumes, and are applicable only in the near-field (downwind distances of no more than a few hundred meters), where the internal turbulence in the plume dominates over the ambient turbulence.

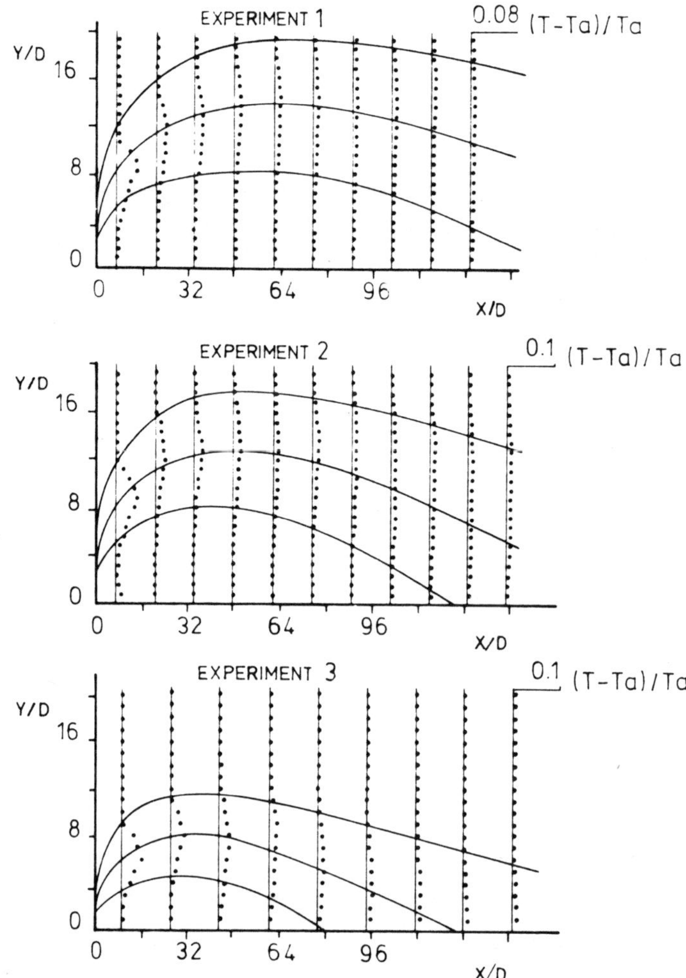


Figure 5-2.  Observed (dots) temperature distributions in a dense gas plume in a wind tunnel.  Solid lines are predicted plume outlines. After Xiao-Yun et al (1986).

Before starting to apply a dense gas jet model, it is important to determine whether the gas is dense enough to have a significant effect on the plume trajectory. If not, then a model for non-buoyant jets (e.g. equation 5-1) can be used. Fay and Zemba (1986) suggest that density effects are important for continuous plumes only if the initial jet Richardson number, $Ri_o$ = $gw_o((\rho_p- \rho_a)/\rho_a))D/u_*^3$, is much greater than one. In this expression, $w_o$ and D are the initial plume speed and diameter, $\rho_p$ and $\rho_a$ are initial plume and ambient density, and $u_*$ is ambient friction velocity (roughly equal to u/15 for nearly-neutral conditions, a wind measurement height of 10m and a roughness length of 1cm). If the $Ri_o >> 1$ criterion is satisfied, then one of the following models should be used. Otherwise one of the standard EPA models for buoyant or neutrally-buoyant plumes can be used.

Hoot et al. (1973) assume that the ambient cross-wind velocity, u, is constant in their dense gas plume model and that the distributions within the plume are constant or "top hat" at any given distance from the source. Also, the specific heats of the gases are assumed equal. These assumptions are also used in the EPA plume rise models. They then define the following basic equations

Entrainment: $\quad d(R^2u_s)/ds = a_1R|u_s - u\cos\theta| + a_2Ru|\sin\theta|$ $\qquad$ (5-6a)

(i.e., changes in volume flux are proportional to differences in velocities inside and outside of the plume around the boundary of the plume)

Horizontal Momentum: $\quad d(\rho_pR^2u^2_s \cos\theta)/ds = \rho_a u d(R^2u_s)/ds$ $\qquad$ (5-6b)

(i.e., the plume is accelerated by drag due to the momentum of the entrained air)

Vertical Momentum: $\quad d(\rho_pR^2u^2_s \sin\theta)/ds = (\rho_a-\rho_p)R^2g$ $\qquad$ (5-6c)

(i.e., vertical momentum changes are due to buoyancy differences)

Mass: $\quad d(\rho_pR^2u_s)/ds = \rho_a\, d(R^2u_s)/ds$ $\qquad$ (5-6d)

(i.e., mass changes are due to entrainment of ambient air)

where R is plume radius, $u_s$ is the plume speed in the direction s, and $\rho_p$ is the plume density. The ambient atmosphere is defined by density $\rho_a$ and speed u. The coordinate system is defined by s, which is the distance along a line parallel to the local plume axis, and $\theta$, which is the angle of this line to the horizontal. The constants $a_1$ and $a_2$ in equation (5-6a) are the constants for along-plume and cross-plume entrainment, which are found to equal 0.09 and 0.90, respectively, based on wind tunnel experiments. Equation (5-6a) is an empirical entrainment assumption that is used to "close" the set of equations, or assure that the number of equations equals the number of variables. Figure 5-3 may help to clarify these definitions. Equations (5-6) are solved analytically by Hoot et al. (1973) in three regions along the plume path. This model yields the following predictions for the maximum initial rise:

$$\Delta h/2R_o = 1.32 \ (w_o/u)^{1/3} \ (\rho_o/\rho_a)^{1/3} \ .$$

$$(w_o{}^2 \rho_o/(2R_o g(\rho_o - \rho_a)))^{1/3} \tag{5-7}$$

where $w_o$, $\rho_o$, and $R_o$ are the initial plume speed, density, and radius. The variables u and $\rho_a$ are the ambient wind speed and density, respectively.

The downwind distance, $x_g$, to plume touchdown is given by the relation:

$$x_g/2R_o = (w_o u \rho_o/(2R_o g(\rho_o - \rho_a))) +$$

$$0.56 ((\Delta h/2R_o)^3 ((2 + h_s/\Delta h)^3 - 1))^{1/2} .$$

$$(u^3 \rho_o/(2R_o g \ w_o(\rho_o - \rho_a)))^{1/2} \tag{5-8}$$

where $h_s/\Delta h$ is the ratio of stack height to maximum plume rise.

The ratio of the maximum concentration C at a given downwind position to the initial concentration $C_o$ is given by:

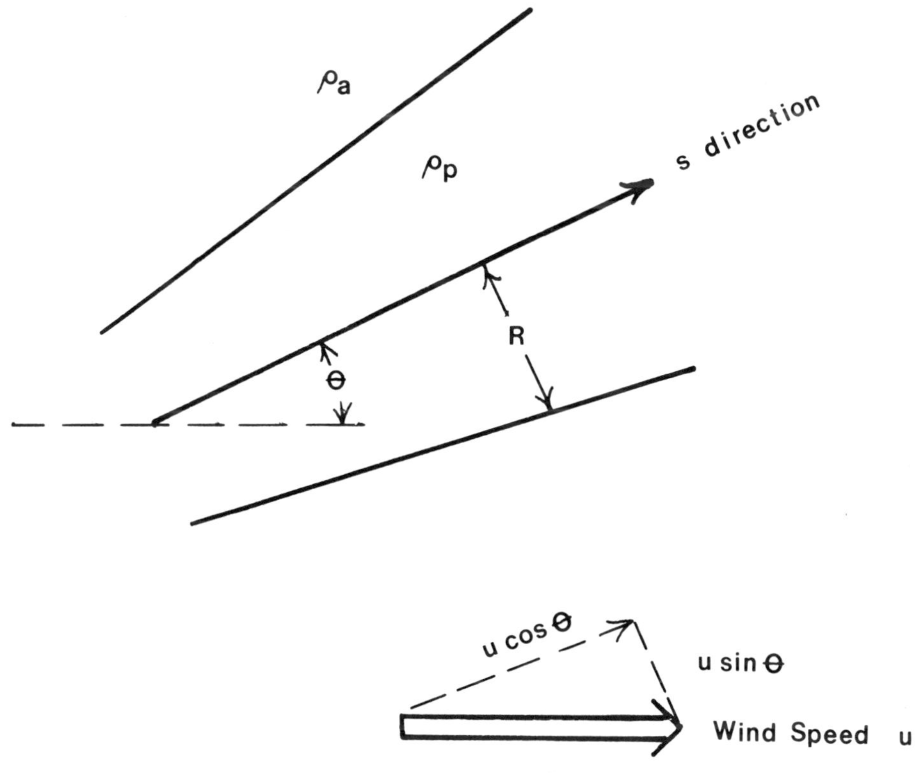


Figure 5-3.  Drawing of plume parameters used in model of Hoot et al (1973). The plume cross-section is circular with radius R.

$$C/C_o = 1.688 \ (w_o/u) \ (\Delta h/2R_o)^{-1.85} \qquad\qquad (5\text{-}9a)$$
$$\text{(at the point of maximum rise)}$$

$$C/C_o = 9.434 \ (w_o/u) \ ((h_s + \Delta h)/2R_o)^{-1.95} \qquad\qquad (5\text{-}9b)$$
$$\text{(at the point } x_g)$$

The plume model by Ooms et al. (1974) differs from the previous model in a few ways. The basic conservation equations are treated in a slightly different way, and the cross-wind distributions of variables are assumed to have a Gaussian or normal shape rather than a top-hat shape. The closure equation is assumed to include a term involving atmospheric turbulence:

$$d(\int_o^{\sqrt{2}R} \rho_p \ u_s \ 2\pi r dr)/ds = 2\pi R \rho_a (a_1 |u_s - u\cos\theta| + a_2 u|\sin\theta|\cos\theta + (\varepsilon R)^{1/3}) \qquad (5\text{-}10)$$

where the variables are defined as before, $a_1$ equals 0.057 and $a_2$ equals 0.50, and $\varepsilon$ is the ambient eddy dissipation rate (typically about $10^{-3}\text{m}^2\text{s}^{-3}$). The plume boundary is assumed to be located at $\sqrt{2}R$ in the integral. Except for the last term this equation is similar to equation (5-6). The set of equations is solved by numerical integration given initial parameters at the border of the "zone of flow establishment" and the "zone of established flow".

### 5.1.2 Dense Gas Releases at Grade

The models for continuous elevated dense gas plumes described above are relevant for routine releases of dense gases. Most accidental releases, however, are from quasi-instantaneous sources near the ground surface. In this case, the models in the first section are not applicable and a new set of models must be developed to cover special situations. In this subsection, so-called box or slab models are reviewed, which assume constant concentrations of plume material at any cross-section. Similarity models are also included in this discussion, since they merely assume a similar form for the concentration distribution at any cross-section.  For example the Gaussian model is a similarity model, since it always assumes that the crosswind distribution has a normal or Gaussian shape given by equation (5-1).  If the density, $\rho_p$, of the cloud is very close to the density of air, then the cloud may be transported and dispersed as if it were buoyant or neutrally-buoyant (i.e. equation 5-1 as found in EPA models may be applied).  A measure of the relative influence of gravity slumping and ambient turbulence on an instantaneous cloud is provided by the source Richardson number:

$$Ri_0 = \text{(Potential energy of cloud)}/\text{(Turbulent energy of environment)} \quad (5-11)$$

Different researchers use slightly different definitions of these two energy components.  For example, Havens and Spicer (1985) use the following definitions for instantaneous and continuous releases:

Instantaneous: $\qquad Ri_0 = g((\rho_p - \rho_a)/\rho_a)V_0^{1/3}/u_*^2$ $\qquad\qquad\qquad$ (5-12)

Continuous: $\qquad Ri_0 = g((\rho_p - \rho_a)/\rho_a)V_0'/uu_*^2 D$ $\qquad\qquad\qquad$ (5-13)

where $V_0$ is the initial volume ($m^3$) of the instantaneous cloud, $V_0'$ is the initial volume flux ($m^3 s^{-1}$) of the continuous plume, and D is the diameter of the release opening.  Fay and Zemba (1986) use $u_*^3$ instead of $uu_*^2$ in equation (5-13).  The friction velocity $u_*$ is employed because it is usually proportional to the turbulent energy components.  There is considerable uncertainty in the value of the "critical" $Ri_0$ separating gravity slumping dominance from ambient turbulence dominance, but it appears to be on the order of ten.  If $Ri_0$ is greater than this critical value, then the dense gas formulations discussed below must be used.

Twenty years ago, the dispersion of dense gas clouds was treated in the same way as the dispersion of inert neutrally-buoyant clouds.  Van Ulden's (1974) experiments with dense gas clouds convinced everyone that something different should be done, since his observed plume dispersion parameters $\sigma_y$ and $\sigma_z$ were factors of four more and less, respectively, than those used for neutrally-buoyant clouds.  This tendency is called dense gas "slumping", and is similar to the flow of a spill of molasses on a table top.  Of all the topics in this book this subject has received the most attention in basic and applied research programs, and has resulted in the largest stack of reprints. Fortunately, there are several excellent review papers on the subject of dense gas slumping (e.g. Blackmore et al. (1982), Raj (1985), Fay (1986), McNaughton et al. (1986)).

It has been observed that some gases with molecular weights less than air (e.g. ammonia) can behave like dense gases due to their cold temperatures and/or the presence of aerosols.

If the sole problem were the slumping of a dense gas cloud at ambient temperature, then the problem could be greatly simplified. However, most dense gas clouds are also cold and contain aerosols, requiring the following thermodynamic phenomena to be modeled:

- Evaporation and condensation of drops and vapor in the cloud.
- Heat exchange with the underlying surface.
- Mass exchange with the underlying surface.
- Radiation flux divergence.
- Chemical changes.

Figure (5-4) characterizes these effects on a single diagram. Note that the complicated thermodynamic aspects relative to the source emissions from the gas/aerosol/liquid spill are not treated here because they have been covered separately in Section 4. Also, in some cases, lower volatility spills may seep into the soil surface before evaporating (Eschenroeder, 1986). Available models generally account for only a few of the phenomena in Figure 5-4.

### Dense Gas Slumping in the Absence of Heat Exchanges

In order to simplify the discussion, it is useful to neglect the five terms in the above list and focus on the dispersion of the dense gas cloud in the absence of heat exchanges. This simplified system has been the subject of many research programs and field and laboratory experiments. It is important to first point out that many gases with molecular weights less than that of air (e.g. ammonia, hydrogen fluoride) will act like dense gases when they are accidentally released to the atmosphere, because of the frequent presence of

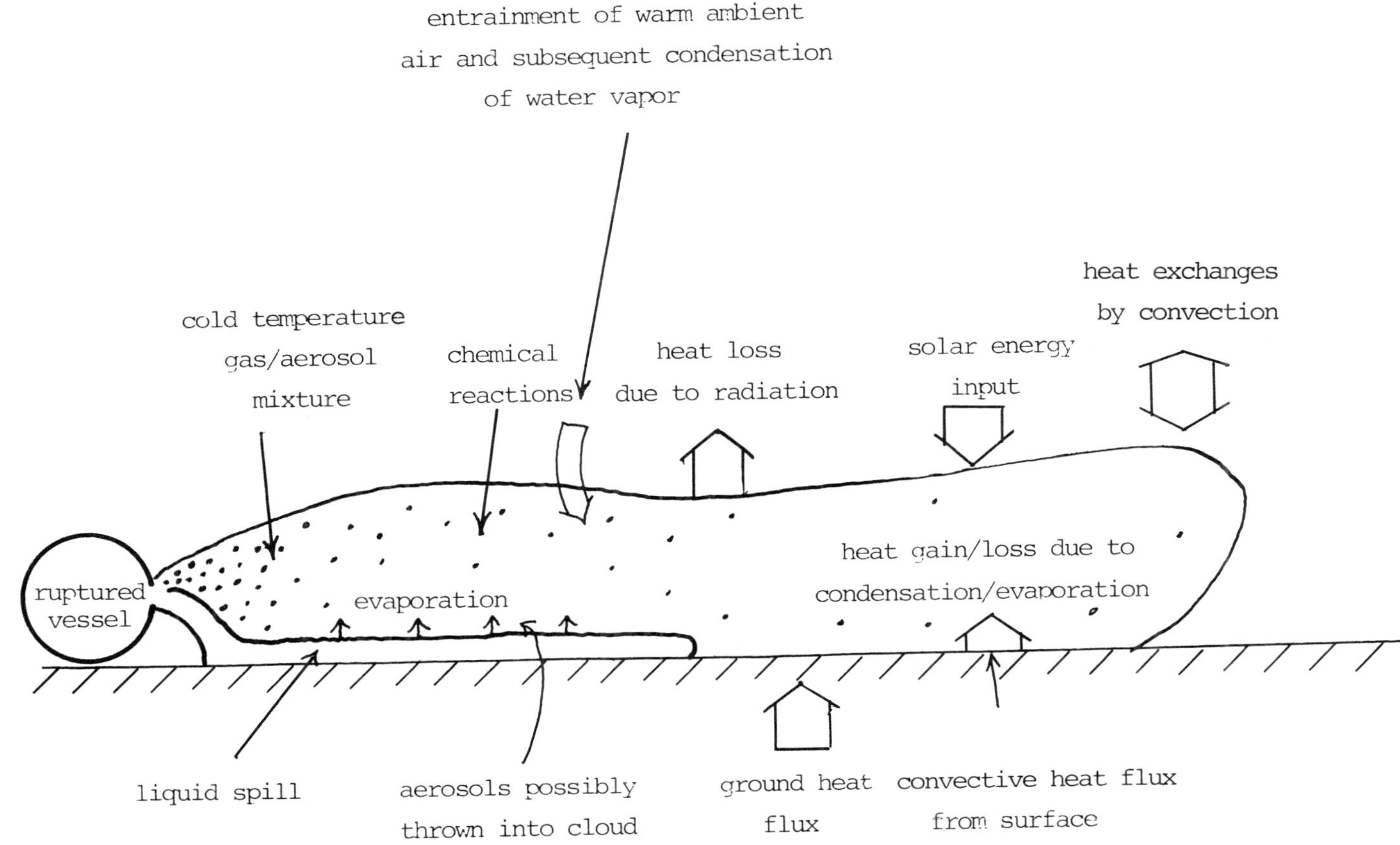


Figure 5-4.  Thermodynamic aspects of a typical hazardous material release.

aerosols in the cloud. If the aerosol drops occupy a small volume fraction of the total mixture, then the density of the two-phase cloud is given by the relation

$$\rho_p = p/RT + \rho_{aer},\qquad\qquad(5\text{-}14)$$

where $\rho_{aer}$ is the mass of aerosol per unit cloud volume (kg m$^{-3}$), p is the atmospheric pressure (in newtons m$^{-2}$, can be assumed constant), T is the absolute temperature ($^O$K) of the cloud, and R, the "gas constant" for the mixture, is given by the formula:

$$R = \sum_{j=1}^{n} \frac{M_j}{M} \frac{R^*}{m_j}\qquad\qquad(5\text{-}15)$$

where $R^*$ is the universal gas constant (8.31436 joules mole$^{-1}$ $^O$K$^{-1}$), $m_j$ is the molecular weight of gas component j, M is the total mass of the gases in the mixture, and $M_j$ is the mass of gas component j. For the purposes of this section, assume that the total mass of aerosol does not change (i.e. there are no heat exchanges and no liquid fall-out).

If the density perturbation in the cloud is important, then the cloud will have a tendency to flow in the horizontal direction rather than diffusing in the vertical direction. The horizontal speed of the edge of the instantaneous cloud, dR/dt, will be proportional to the square root of the numerator in equation (5-12):

$$dR/dt \propto (g((\rho_p-\rho_a)/\rho_a)V^{1/3})^{1/2}\qquad\qquad(5\text{-}16)$$

The constant of proportionality is found to equal about unity. If there are no heat additions to the cloud, then the buoyant force on the right side of equation (5-16) will remain constant (Fay 1986) and the following well-known formula for the cloud radius will result:

$$R^2 = R_o{}^2 + 2(g((\rho_p-\rho_a)/\rho_a)_o V_o{}^{1/3})^{1/2}t\qquad\qquad(5\text{-}17)$$

This formula has been validated with both field and laboratory data, and an example of observations from the Thorney Island trials is given in Figure 5-5. The source was an instantaneously released volume (about $2000m^3$) of freon gas. Except for the "flow establishment" zone at small times, the data clearly justify the use of equation (5-17). Note that the radius R of the slumping cylinder defined here is horizontal, where the radius R of the elevated plume defined in Figure 5-3 is perpendicular to the plume axis.

In the case of a continuous release at the ground, the increased spread is felt mainly in the lateral or cross-wind direction and equation (5-16) must be modified slightly to reflect the fact that $v^{1/3}$ is basically the depth of the plume, h. Furthermore the time derivative must be converted to a distance derivative (Raj 1985) yielding:

$$u_e \ dR/dx \ \propto (g((\rho_p-\rho_a)/\rho_a)h)^{1/2} \tag{5-18}$$

To solve this equation the effective velocity of the plume, $u_e$, must be known. Some researchers assume that the effective wind speed is equal to the wind speed at some fraction (say 0.4) of the plume depth. Others assume that the effective wind speed is equal to the observed wind speed times the term $(1 - V_o'/V')$, which accounts for the inertia of the initial release. It is simplest to assume that $u_e$ equals the wind speed at some standard measurement height, say 2m. Again, if there are not heat additions to the plume, the buoyancy flux, $gu_ehR(\rho_p-\rho_a)/\rho_a$, will be conserved, yielding the following solution (Raj 1985) for a continuous dense gas plume:

$$R = R_o \ (1 + 1.5(gh_o((\rho_p-\rho_a)_o/\rho_a))^{1/2} \ x/u_eR_o)^{2/3} \tag{5-19}$$

Note that the cloud width grows at a faster rate for the continuous plume than for the instantaneous cloud, since there is one less degree of freedom for the continuous plume.

The procedures described above will permit the horizontal dimension of the dense gas cloud to be estimated. But in order to calculate concentrations in the cloud, its vertical depth, h, must also be predicted. Together, they

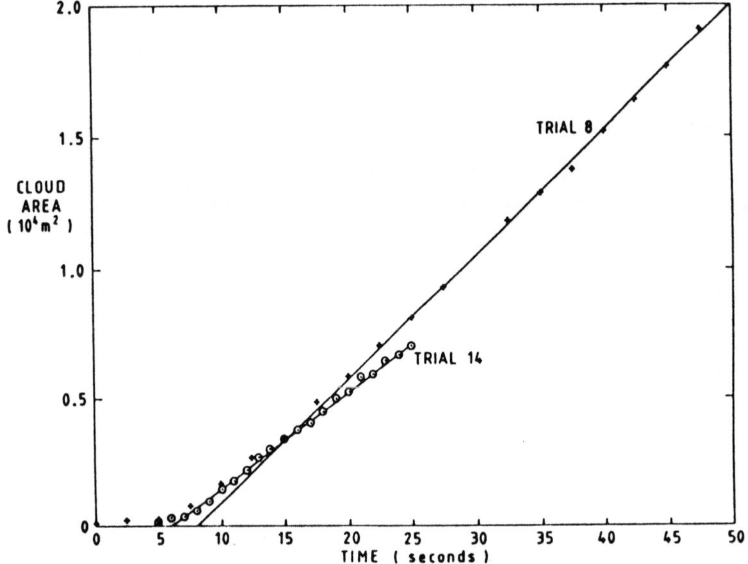


Figure 5-5.  Area of dense gas cloud versus time for two Thorney Island trials (Brighton et al 1985).

64

yield the dilution of the initial cloud or plume.  For constant gas
concentrations (i.e. a top hat distribution) at a given travel time or downwind
distance, the decrease in concentrations are given by the formulas:

Instantaneous $\qquad C/C_0 = V_0/V = (h_0 R_0^2)/(hR^2)$ $\qquad\qquad$ (5-20)

Continuous $\qquad\qquad C/C_0 = V_0'/V' = V_0'/u_e hR$ $\qquad\qquad\qquad$ (5-21)

These two ratios are determined by the entrainment rate of ambient fluid into
the cloud due to turbulence at its edges.  Rather than predicting h directly,
the models generally predict the ratio $V_0/V$ or $V_0'/V'$.  There are a great
variety of methods available to parameterize the entrainment rate, and thus
obtain solutions to equations (5-20) and (5-21), and these will be more
completely reviewed in a later subsection.  However, it is evident from the
Thorney Island observations that the dilution at that site is proceeding
independent of initial Richardson number $Ri_0$ and wind speed (see Figure 5-6,
from McQuaid 1985).  This result suggests that the problem can be treated using
simple dimensional analysis.

Equation (5-16) is an example of a result that can be obtained solely
through dimensional analysis.  It can be written in the form

$$RdR/dt = E_0^{1/2} \qquad\qquad\qquad (5-22)$$

where $E_0 = (g/\rho_a)(\rho_0 - \rho_a)V_0$ is the initial potential energy $(m^4 s^{-2})$ for an
instantaneous release.  Dimensional analysis has been used by many researchers
to derive useful results in geophysical research.  Briggs (1969) derived
several plume rise formulas by this procedure which form the basis of current
EPA UNAMAP6 regulatory models.  According to the basic principles of
dimensional analysis (the so-called Buckingham Pi theorem), if there is a set
of n independent parameters and variables important to a problem and these
parameters have m independent dimensions, then n-m independent dimensionless
parameters can be formed.  Since R, t and $E_0$ all employ length and time units
then n=3, m=2, and equation (5-22) is the only possible result.  However,

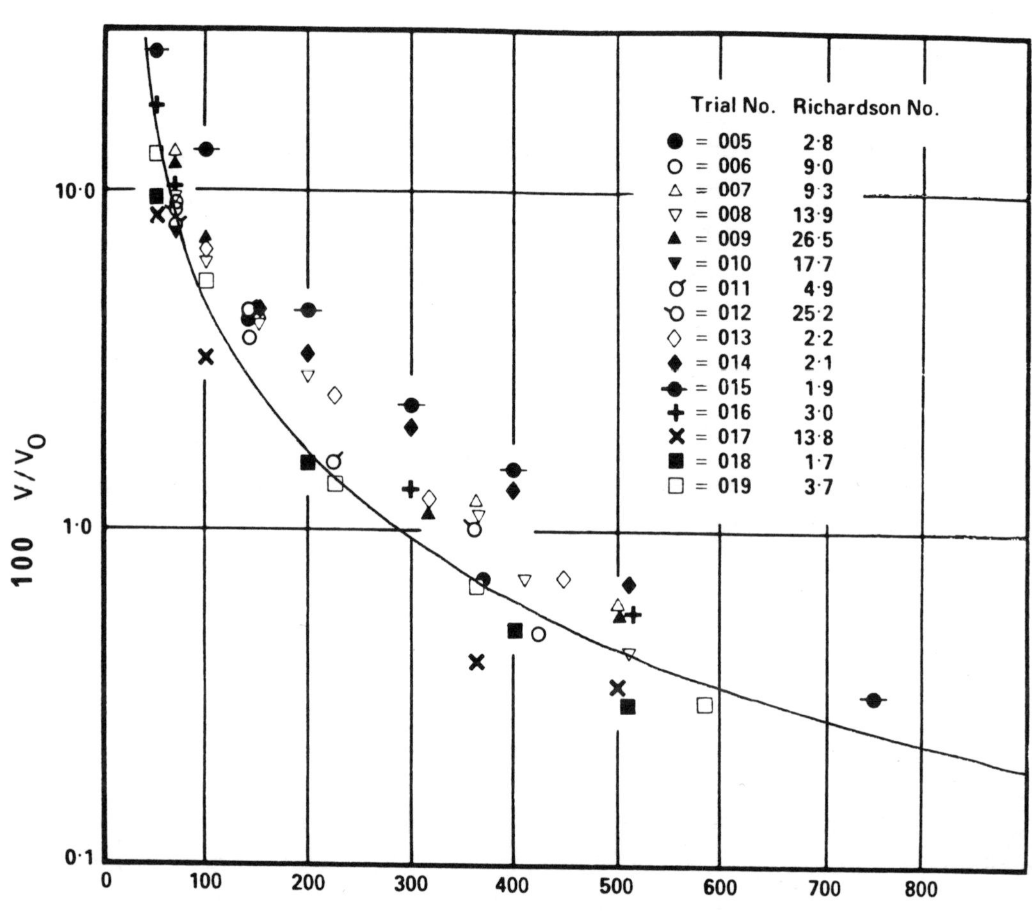


Figure 5-6.  Thorney Island observations of dilution as a function of downwind
distance.  The predicted curve from equation 5-24 is drawn.
McQuaid (1985) presented the original figure minus the predicted
curve.

dimensional analysis does not give the proportionality constant or the functional forms in which the groups appear in the equations. The relationship happens to be linear in equation (5-22) with a constant of proportionality of unity.

The dilution ratio $V/V_O$ can also be estimated by means of functional dependencies derived from dimensional analysis combined with empirical data. This ratio is independent of $u_*$ initially and therefore must also be independent of $PE_O$, since there is no other parameter in the set of initial parameters that includes dimensions of time. Thus n equals 3, m equals 1, and we can write:

$$V/V_O = f(x/V_O^{1/3}) \tag{5-23}$$

where f indicates a generalized dimensionless function. The function f can be defined by plotting field observations of $V/V_O$ versus $x/V_O^{1/3}$, as shown in Figure 5-6 for the Thorney Island data. The initial volume, $V_O$, in these trials is about $2000m^3$ in all cases. Instantaneous releases of Freon gas took place in these experiments, with wind speeds ranging from about 1 m/s to 8 m/s and $(\rho_O - \rho_a)/\rho_a$ ranging from about 0.4 to 4. The line on the figure is given by the formula:

$$V/V_O = (x/V_O^{1/3})^{1.5} \tag{5-24}$$

valid for $x \geq V_O^{1/3}$

The scatter of the points on the figure is relatively small as atmospheric data go and tend to verify the assumption that the ratio $V/V_O$ is not a function of $Ri_O$ or wind speed. All the points are within a factor of two of the line given by equation (5-24). At greater distances downwind, it is expected that the initial Richardson number $Ri_O$ should be included in equation (5-23), since ambient turbulence eventually dominates the dispersion.

This simple result verifies the results of dense gas model comparisons reported by other researchers, such as Havens and Spicer (1985) and McNaughton et al. (1986). These other models seem to compare quite well with the Thorney

Island data as long as the gross entrainment $(1/V)dV/dt$ is properly parameterized. However, it is required that the data be in the range where gravity slumping is dominant. If it is assumed that gravity slumping ceases and ambient dispersion takes over when the density perturbation $(\rho_o - \rho_a)/\rho_a$ drops below some limit $\varepsilon$, then this occurs at a distance $x_f$ given by

$$\varepsilon = (x_f/V_o^{1/3})^{-1.5} \ (E_o/gV_o)$$

$$\text{or} \quad x_f = E_o^{2/3} \ V_o^{-1/3}/(g\varepsilon)^{2/3} \tag{5-25}$$

$$= 21.5 \ E_o^{2/3}V_o^{-1/3}/g^{2/3} \qquad ( =.01)$$
$$= 100 \ E_o^{2/3}V_o^{-1/3}/g^{2/3} \qquad ( =.001)$$

For the Thorney Island experiments, where $V_o = 2000m^3$ and $E_o = 20000m^4/s^2$, $x_f$ equals 271m and 1260m for $\varepsilon$ equal to .01 and .001, respectively. By this criterion, most of the data in Figure 5-6 are within the range of distances where gravity slumping is important.

Slightly different results regarding $x_f$ are obtained if it is assumed that the transition occurs when the local $Ri = g((\rho-\rho_a)/\rho_a)V^{1/3}/u_*^2$ drops below a critical value. This gives the formula

$$x_f = E_o/(V_o^{1/3}u_*^2Ri_c)$$
$$= 15870m \qquad (Ri_c = 1)$$
$$= 1587m \qquad (Ri_c = 10) \tag{5-26}$$

Havens and Spicer (1985) suggest that the $Ri_c = 10$ criterion is more valid, which would suggest that all the observations in Figure 5-6 are dominated by dense gas slumping.

This analysis shows how simple formulas based on dimensional analysis can explain many observations of dense gas dispersion. It is not necessary to specifically model the detailed flows inside the gas cloud, since the gross entrainment integrated over the cloud is important. The Thorney Island observations verify this straightforward approach.

## Dense Gas Slumping in the Presence of Heat Exchanges

The derivations in the previous subsection were for a gas cloud that did not exchange heat with its environment or generate heat internally through latent heat processes or chemical conversions.  In practical situations, the atmosphere will attempt to eliminate any temperature gradients that are initially present.  For example, a cold cloud resulting from a spill from a pressurized tank will be warmed by heat transfer from the underlying ground or sea surface, by entrainment of warm ambient air, and by latent heat released by condensation of ambient water vapor.  On the other hand it could be cooled by the emission of long wave radiation from the upper surface of its aerosol cloud or by evaporation of aerosol droplets.  These internal heat changes can also affect the dispersion of the cloud.  The characteristics of many of these heat transfer parameters are discussed below:

- **Heat transfer from below the surface**:  The rate at which heat can be added to the cold cloud from the surface is limited by the thermal conductivity of the substrate.  Obviously water is a more efficient conductor than loose, dry gravel, although a complication can occur with water if the surface freezes and the conductivity decreases.

  The temperature at a given level in the ground will continually decrease as heat is drawn from the ground, and the energy equation can be simplified such that the local rate of change of temperature with time is given by the Fourier heat conduction law:

$$\partial T/\partial t = (k/c\rho)\partial^2 T/\partial z^2 \tag{5-27}$$

  where $k$, $c$, and $\rho$ are the thermal conductivity ($wm^{-1}{}^{o}K^{-1}$), specific heat ($j\ kg^{-1}{}^{o}K$), and density ($kg\ m^{-3}$) of the soil.  Boundary conditions are that $T$ approaches a constant at great depths and the flux $-k\partial T/\partial z$ approaches the sensible plus latent heat fluxes in the air from the ground to the cloud (in the absence of radiative effects).  The thermal conductivity $k$ is observed to range over several orders of magnitude, and values for special surfaces are given by Oke (1978). For example the Shell SPILLS models uses values of 0.32 and 2.21 $Wm^{-1}\,{}^{o}K^{-1}$ for dry and wet soil, respectively.

- <u>Heat transfer from the surface to the cloud</u>: If the ground surface is dry, then all heat transfer from the surface to the cloud is due to sensible heat flux, $H_s$ (watts $m^{-2}$) Atmospheric boundary layer theory provides equations for parameterizing this heat flux under normal conditions in the absence of a cold cloud (Panofsky and Dutton 1984). If a cold cloud is present then the sensible heat flux is due mainly to natural convection, and if the temperature difference between the ground and the cloud is large (say $10\,^{o}C$ or greater) then heat transfer is by free convection (i.e. independent of the wind speed). Wu and Schroy (1979) suggest the empirical formula:

$$H_s = h_c(T_s - T_a) \tag{5-28}$$

where $T_s$ is the ground surface temperature, $T_a$ is the cloud temperature, and $h_c$ is the convection coefficient, assumed equal to 4.5 watts/$m^2\,{}^{o}K$.

If there is a liquid at the surface (e.g., water or spilled chemical), then the latent heat of evaporation is important. Each gram of liquid that evaporates at the surface will take heat from the substrate. By analogy with equation (5-28), the "free-evaporation" is given by the formula:

$$H_E = h_e L\,(\rho_{gs} - \rho_g) \tag{5-29}$$

where $\rho_g$ is the gas density in the cloud, L is the latent heat of vaporization for a particular liquid, and $h_e$ is a transfer coefficient equal to about 1 m/s. It is assumed that $\rho_{gs}$ is the saturation density of the gas at the surface temperature. These formulas are valid for single components.

**5. Transport and Dispersion Models**

- <u>Heat transfer by evaporation/condensation processes within the cloud</u>: If the cloud initially contains an aerosol with density $\rho_{aer}$, then that aerosol will take from the cloud an amount of heat (per unit volume of the cloud) equal to $L\rho_{aer}$ when it all evaporates. The rate of evaporation depends on the temperature variation within the dense gas cloud. This mechanism is responsible for the maintenance of the density difference in aerosol clouds whose molecular weight is less than that of air (e.g., ammonia). The evaporation of aerosols is said to have a "refrigerator" effect.

As ambient air is entrained by the cold, dense cloud, the ambient air is cooled. Depending on the water vapor mixing ratio in the ambient air, pure water drops can be condensed from the entrained air. As these drops condense they release their latent heat of vaporization to the cloud. The maximum amount released per unit volume is equal to the latent heat of vaporization times the difference between the saturation density of water vapor in the cloud and the actual density of water vapor in the ambient air. These evaporation/condensation formulas are likely to be overestimates, since the heat exchanged during the phase change alters the temperature of the cloud.

- <u>Temperature change by dilution</u>: The ambient atmosphere and the slumping cloud are nearly always turbulent and thus are very diffusive. By the time a cloud of hazardous materials is transported 1 km downwind, its volume will have increased by several orders of magnitude due to entrainment of ambient air. Assuming no other heat inputs and equal molar heat capacities, the new temperature of the mixture is given by the formula:

$$\rho V T \text{ (mixture)} = \rho_a (V-V_o) T_a + \rho_o V_o T_o \qquad (5\text{-}30)$$

where $\rho$ and $V$ are the molar density and the volume of the mixture, and subscript o refers to the initial gas release. While this equation does not describe a heat exchange itself, it strongly influences the heat exchange processes because of the temperature change of the mixture. Methods of estimating the dilution $V/V_o$ are reviewed at several places in this section.

71

Other heat exchange processes that have not been treated here, such as radiative flux divergences and releases of heat during chemical reactions, are generally small, but may become important for special problems.  More research needs to be performed in these areas.

The energy exchange processes discussed above are summarized in Figure 5-7.

A Comparison of Entrainment Assumptions in Simple Models

Before discussing specific aspects of dense gas cloud dispersion models, it is helpful to develop a framework of basic equations to which many of the models can be referred.  So-called slab or box models are the most prolific because they are simple to understand and use, and they agree with the limited data that are available as well as more complex models.  They generally assume that parameters such as temperature or density are uniform across the cloud after a given travel time or downwind distance.  Webber (1983) published a monograph in which the basic framework of slab or box models is developed and used to compare several models.  Raj (1985) and Fay (1986) also discuss these generalized equations.

Webber (1983) assumes that the cloud is cylindrical with depth h and horizontal radius R, and that the heat exchanges discussed in the preceding sections are unimportant.  The following equations are derived:

- Conservation of Buoyancy or Potential Energy

$$E = g((\rho - \rho_a)/\rho_a)hR^2 = E_o \tag{5-31}$$

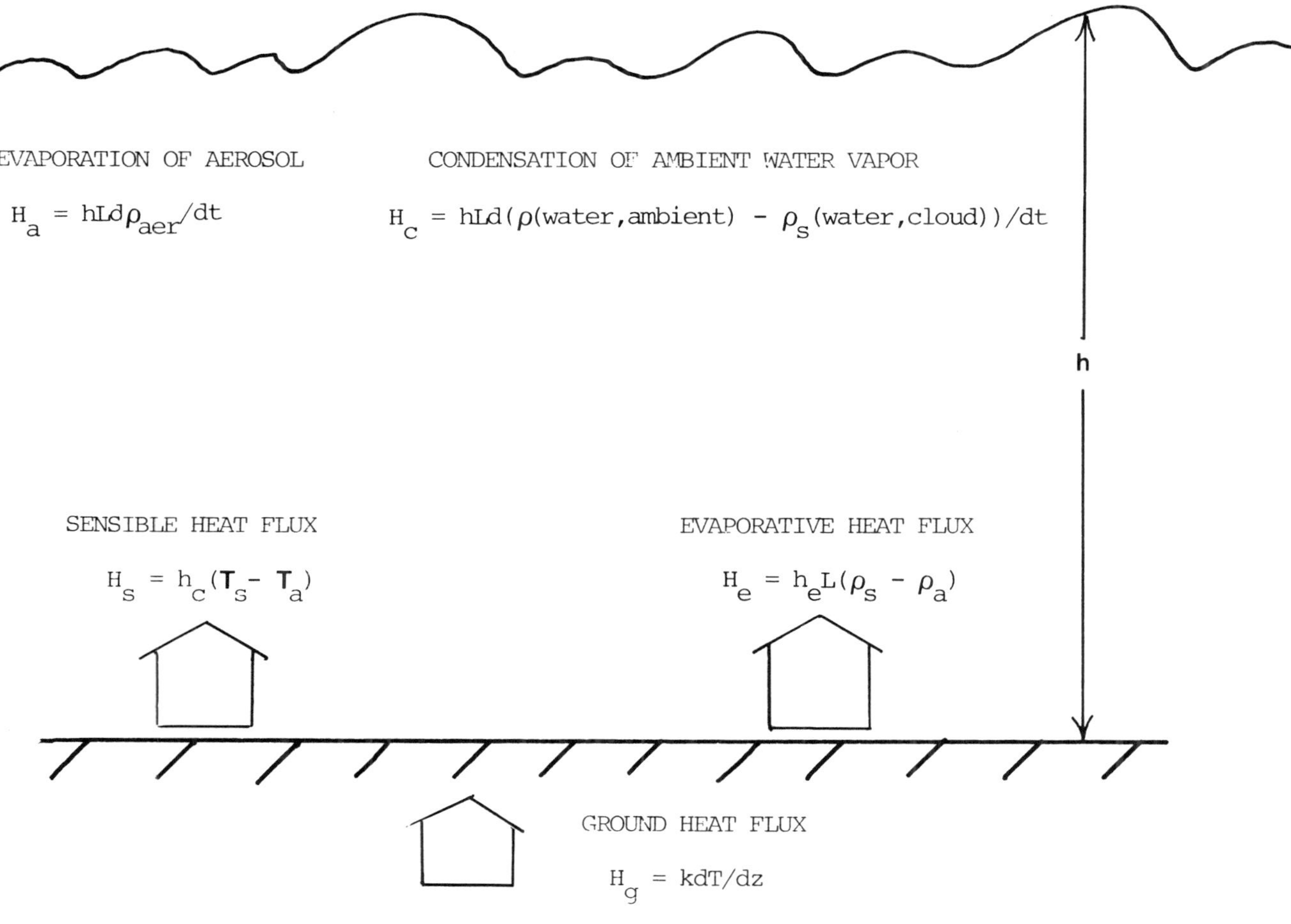


Figure 5-7. Components of heat exchange in a dense gas cloud of depth h. The density $\rho$ and latent heat L refer to the particular gas or liquid being described.

where $E_o$ is the initial value.

- Air Entrainment or Closure

$$dV/dt = (\pi R^2) U_T + (2\pi Rh) U_E \qquad (5\text{-}32)$$

where $V = \pi hR^2$ and $U_T$ and $U_E$ are top and edge entrainment velocities. This closure prescription varies greatly from model to model, although it is clear that it is unimportant how the components are prescribed, as long as the total $dV/dt$ is well-simulated.

- Gravity Front Equation

$$dR/dt = U_f \propto (gh((\rho - \rho_a)/\rho_a))^{1/2} \qquad (5\text{-}33)$$

where the constant of proportionality is found to equal about unity. Some models use $\rho$ in the denominator rather than $\rho_a$. This equation has also been discussed in previous subsections (see equations 5-16 and 5-22).

This set of equations is used for the analysis of many problems where gravity fronts are important, such as the spread of salt water tongues in fresh water bodies and the spread of the downburst from a thunderstorm. The all-important assumptions for the entrainment velocities $U_T$ and $U_E$ are listed for a few models in Table 5-1, which is a reduced version of a table published by Raj (1985). The purpose of this table is to illustrate the remarkable variability in these assumptions. But it must be realized that the division between $U_T$ and $U_E$ is irrelevant as long as $dV/dt$ is simulated. As shown in equation (5-24), the following variation is indicated by the Thorney Island data:

$$dV/dt = 1.5 \, u_e V_o^{2/3} \, (x/V_o^{1/3})^{1/2}, \qquad (5\text{-}34)$$

where $u_e$ is the effective transport velocity of the cloud. Once $V/V_o$ is greater than about ten (i.e. at a distance of about two source radii downstream, then the initial inertia of the dense gas is unimportant and $u_e$ is equal to the wind speed at a height of about 0.4h (Fay 1986).

5. **Transport and Dispersion Models**

Table 5-1

Comparison of Different Box Model Entrainment Assumptions (after Raj 1985)

| Model: | Van Ulden 1974 | Germeles & Drake 1975 | Picknett 1981 | Fryer & Kaiser Cox & Carpenter 1979 | Eidsvik 1980 | Fay & Ranck 1983 |
|---|---|---|---|---|---|---|
| Edge Entrainment | $\alpha U_f$ | 0 | $\alpha U_f$ | $\alpha U_f$ | $\alpha U_f^2/U_f(0)$ | 0 |
| Top Entrainment | 0 | $\beta U_f$ | $\beta U_*/Ri$ | $\beta U_*/Ri$ | $\beta v_e/(\alpha + Ri)$ | $\beta u_*/(\alpha^2 + Ri^2)^{1/2}$ |

Note:  $\alpha$ and $\beta$ are constants; Ri is the source Richardson number, $v_e$ is the resultant of $U_f$ and the wind speed.

5.1.3  Transition to Neutrally-Buoyant Plume Transport and Dispersion

At some travel time or downwind distance, the cloud or plume ceases to be
influenced by the density perturbation and subsequent transport and dispersion
proceeds as if the plume were neutrally buoyant.  A complete three dimensional
numerical code could handle all situations and the transition would be implicit
in the code.  Similarly, entrainment models have been written that include
components for density effects and for entrainment due to ambient turbulence
(e.g., Eidsvik, 1980).  Most of the models in Table 5-1 have arbitrary
assumptions for the transition point.

Equation 5-25 provides one example of this transition assumption:
slumping ceases when the density perturbation $(\rho_o - \rho_a)/\rho_a$ drops below some
limit  (e.g. .01 or .001).  Equation 5-26 provides another criterion:  the
local $Ri = g(\rho - \rho_a)/\rho_a)V^{1/3}/u^2$ drops below a critical value (e.g. 1 or 10).
Some examples of other criteria in Table 5-1 are listed below:

| | | |
|---|---|---|
| $U_f = 2u_*$ | van Ulden (1974) | (5-35a) |
| $U_f = u$ | Germeles and Drake (1975) | (5-35b) |
| $U_f = d\sigma_y/dt$ | Cox and Carpenter (1979) | (5-35c) |
| $U_f = 0.39u_*$ | Eidsvik, (1980) | (5-35d) |

These criteria cover a wide range, from $U_f = 0.02u$ to $U_f = u$ (assuming $u_* \sim$
$u/15$).  The Germeles and Drake (1975) criterion is so restrictive that the
dense gas slumping phase exists for only a very short time, if at all, for
typical source strengths.  Raj (1985) and Fay (1986) point out that most
researchers now recognize that $U_f$ is on the order of $u_*$, the friction velocity,
which is the square root of the surface momentum flux divided by the density.
The obvious uncertainty in the transition criterion leads to considerable
uncertainty in concentration calculations at that distance.

At the transition point, most box or slab models change over to a Gaussian
plume or puff model (given in equation 5-1).  This transition can be
accomplished by means of a virtual source procedure, where for a puff it can be
assumed that the dispersion paramaters $\sigma_z$, $\sigma_y$, and $\sigma_x$ are given by the
relations:

$$\sigma_z = h/\sqrt{2} \tag{5-36a}$$

$$\sigma_y = R/\sqrt{2} \tag{5-36b}$$

$$\sigma_x = \sigma_y \text{ (if instantaneous source)} \tag{5-36c}$$

It is assumed that a virtual source exists some distance $x_v$ upstream from the transition point, such that the maximum concentration in the Gaussian plume would equal the value calculated by the box or slab model. The lateral and vertical dispersion parameters, $\sigma_y$ and $\sigma_z$, in a neutrally-buoyant dispersion model (e.g. the EPA UNAMAP models) can be written as functions of x or t and inverted to solve for the virtual source distance, $x_v$, or time, $t_v$. For example, if $\sigma_y = 0.1x$, then $x_v = 10R/\sqrt{2}$.

The dispersion parameters for instantaneous sources (duration of release less than averaging time or travel time from source to receptor) are known to be different than those for time-averaged plumes from continuous sources (Hanna et al., 1982). But since there are many more data available for continuous sources, it is customary to use those values for all types of sources. The equations in Table 5-2 can be used to calculate virtual distances and dispersion of the neutrally-buoyant gas. Stability class A is very unstable, D is neutral, and F is very stable. The dispersion rates decrease in general as atmospheric stability increases. Table 5-3 provides a mechanism for estimating the stability class.

Table 5.2 Formulas Recommended by
Hanna, Briggs, and Hosker (1982) for $\sigma_y(x)$
and $\sigma_z(x)$ (100m < x < 10km) for
Continuous Plumes Averaged over about
Ten Minutes to One Hour

| Stability Class | $\sigma_y$ (m) | $\sigma_z$ (m) |
|---|---|---|
| A | $0.22x\,(1 + 0.0001x)^{-1/2}$ | $0.20x$ |
| B | $0.16x\,(1 + 0.0001x)^{-1/2}$ | $0.12x$ |
| C | $0.11x\,(1 + 0.0001x)^{-1/2}$ | $0.08x\,(1 + 0.0002x)^{-1/2}$ |
| D | $0.08x\,(1 + 0.0001x)^{-1/2}$ | $0.06x\,(1 + 0.0015x)^{-1/2}$ |
| E | $0.06x\,(1 + 0.0001x)^{-1/2}$ | $0.03x\,(1 + 0.0003x)^{-1}$ |
| F | $0.04x\,(1 + 0.0001x)^{-1/2}$ | $0.016x\,(1 + 0.0003x)^{-1}$ |

Table 5-3

Meteorological Conditions Defining Pasquill Stability Classes[*]

A: Extremely unstable conditions      D: Neutral conditions[+]
B: Moderately unstable conditions     E: Slightly stable conditions
C: Slightly unstable conditions       F: Moderately stable conditions

| Surface wind speed, m/sec | Daytime insolation[+++] | | | Nighttime Conditions[++] | |
| --- | --- | --- | --- | --- | --- |
| | Strong | Moderate | Slight | Thin overcast or >4/8 low cloud | <3/8 cloudiness |
| <2 | A | A-B | B | | |
| 2-3 | A-B | B | C | E | F |
| 3-4 | B | B-C | C | D | E |
| 4-6 | C | C-D | D | D | D |
| >6 | C | D | D | D | D |

[*] From F.A. Gifford, Turbulent Diffusion-Typing Schemes: A Review, Nucl. Saf., 17(1):71(1976).
[+] Applicable to heavy overcast day or night.
[++] The degree of cloudiness is defined as that fraction of the sky above the local apparent horizon that is covered by clouds.
[+++] Appropriate insolation categories may be determined through the use of sky cover and solar elevation information as follows:

| Sky Cover | Solar Elevation Angle > 60° | Solar Elevation Angle ≤ 60° But > 35° | Solar Elevation Angle ≤ 35° But > 15° |
| --- | --- | --- | --- |
| 4/8 or Less or Any Amount of High Thin Clouds | Strong | Moderate | Slight |
| 5/8 to 7/8 Middle Clouds (7000 feet to 16,000 foot base) | Moderate | Slight | Slight |
| 5/8 to 7/8 Low Clouds (less than 7000 foot base) | Slight | Slight | Slight |

Ground level concentrations due to dispersion beyond the transition point can be calculated using the Gaussian puff or plume equation:

$$\text{Puff} \quad C/Q_i = (\sqrt{2}\pi^{3/2}\sigma_x\sigma_y\sigma_z)^{-1}\exp(-(x-x_o)^2/2\sigma_x^2)\exp(-(y-y_o)^2/2\sigma_y^2)\cdot$$
$$\exp(-z_e^2/2\sigma_z^2) \qquad (5\text{-}37)$$

$$\text{Plume} \quad C/Q = (\pi u\sigma_y\sigma_z)^{-1}\exp(-(y-y_o)^2/2\sigma_y^2)\exp(-z_e^2/2\sigma_z^2) \qquad (5\text{-}38)$$

where $Q_i$ (g) is the instantaneous source strength and $Q(\text{gs}^{-1})$ is the continuous source strength. Other modeling approaches for dispersion in this region are discussed in the next subsection.

5.1.4 Modeling Approaches other than Box or Slab Models

Some modelers distinguish between box or slab models and so-called "similarity" models. For example, the Gaussian model (equations 5-37 and 5-32) is a similarity model because the crosswind distribution is always similar (Gaussian). In this review, however, similarity models are considered to be in the same class as box or slab models. Three dimensional numerical models and physical (laboratory) models have also been applied to calculate vapor cloud dispersion, and these approaches are reviewed below.

Three Dimensional Models

Simple box, slab, or similarity models provide great insight into the physical processes important for the transport and dispersion of hazardous materials. Furthermore, they are shown to agree fairly well with the limited data bases. Yet it is obvious that there is much structure within the plume or cloud that is not being accounted for by the model. Several models have been developed that do account for varying gradients and flux divergences within the plume. Most of these are based on solutions to the three-dimensional, time-dependent conservation equations, and often require a main-frame super-computer. A set of governing equations used by Ermak and Chan (1985) is listed below:

o     Equations of Motion

$$\partial(\rho_p\underline{u})/\partial t + \rho_p\underline{u}\cdot\nabla\underline{u} = -\nabla(p-p_a) + \nabla\cdot(\rho_p\underline{\underline{K}}^m\cdot\nabla\underline{u}) + (\rho_p-\rho_a)\underline{g} \qquad (5\text{-}39)$$

o     Equation of Mass Continuity

$$\nabla\cdot(\rho_p\underline{u}) = 0 \qquad (5\text{-}40)$$

o     Equation of State

$$\rho_p = pm/R^*T = p/(R^*T((\rho_pC/m_j)+(1-\rho_pC/m_a))) \qquad (5\text{-}41)$$

o     Mass Conservation of Pollutant

$$\partial\rho_pC/\partial t + \underline{u}\cdot\nabla\rho_pC = (1/\rho_p)\nabla\cdot(\rho\underline{\underline{K}}^C\cdot\nabla\rho_pC) \qquad (5\text{-}42)$$

o     Conservation of Enthalpy

$$\partial\theta/\partial t + \underline{u}\cdot\nabla\theta = (1/\rho_pc_p)\nabla\cdot(\rho_pc_p\underline{\underline{K}}^\theta\cdot\nabla\theta) \qquad (5\text{-}43)$$
$$+((c_{pn}-c_{pa})/c_p)(\underline{\underline{K}}^C\cdot\nabla\rho_pC)\cdot$$

where vector (underlines) and tensor (double underline) notation is used and the dot represents the dot product (e.g. $\underline{u}\cdot\nabla C = u\partial C/\partial x + v\partial C/\partial y + w\partial C/\partial z$).   The following definitions are different from those appearing earlier:

$C$     = Concentration of hazardous gas in the cloud $(kg/m^3)$

$c_p$     = Specific heat of the mixture $(j\ kg^{-1}\ {}^oK^{-1})$

$c_{pn}$     = Specific heat of the hazardous gas $(j\ kg^{-1}\ {}^oK^{-1})$

$c_{pa}$     = Specific heat of the air $(j\ kg^{-1}\ {}^oK^{-1})$

$\nabla$     = Del operator $(\underline{i}\partial/\partial x + \underline{j}\partial/\partial y + \underline{k}\partial/\partial z)$

$\underline{u}$     = $\underline{i}u + \underline{j}v + \underline{k}w\ (m\ s^{-1})$

$m_j$     = Molecular weight of the hazardous gas

$m_a$     = Molecular weight of air

$\underline{\underline{K}}$     = Eddy diffusivity coefficient $(m^2s^{-1})$, which is a function of local turbulence and stability

p = pressure in the cloud and $p_a$ is the atmospheric pressure in the absence of the cloud (newtons/$m^2$).

Superscripts on the $\underline{\underline{K}}$'s refer to the values appropriate to the variable in question.

Note that, despite the complexity of these equations, interphase heat exchange processes (e.g. evaporation of droplets) within the cloud are neglected. The above set of equations is solved by computer over a three-dimensional set of grid points, given appropriate initial and boundary conditions. Any set of equations describing dispersion also requires a closure hypothesis. The box or slab models used an entrainment assumption to close the system. Similarly, the set of equations (5-39) through (5-43) require specification of the K coefficients to close the system. These coefficients are very uncertain for hazardous gas systems and probably are the major contributors to errors in the three dimensional numerical models. Chan and Ermak (1985) use the following formulas for the $K^{m'}$s (note that only the diagonal terms of the tensor $\underline{\underline{K}}$ are assumed to be non-zero):

$$\text{Vertical } K^m_z = 0.4((u_*z)^2 + (w_*1)^2)^{1/2}/\phi(Ri) \qquad (5\text{-}44)$$

$$\text{Horizontal } K^m_y = 0.4Bu_*z/\phi(Ri) \qquad (5\text{-}45)$$

$$\text{where } \phi(Ri) = 1 + 5\ Ri \qquad (Ri \geq 0) \qquad (5\text{-}46)$$

$$\phi(Ri) = (1-16\ Ri)^{-1/4} \text{ for momentum} \qquad (Ri<0) \qquad (5\text{-}47)$$

$$\phi(Ri) = (1-16\ Ri)^{-1/2} \text{ for energy and species } (Ri<0) \qquad (5\text{-}48)$$

$$\text{and } Ri = u_*^2\ Ri_a/(u_*^2 + w_*^2) + 0.05((\rho_p - \rho_a)/\rho_p)g1/(u_*^2 + w_*^2) \qquad (5\text{-}49)$$

where B is an adjustable parameter and 1 is the scale length of the cloud (equal approximately to the cloud depth). The convective velocity $w_*$ is due to ground heating of the cloud and is proportional to the cube root of the heat flux times the mixing depth. The local Richardson number, Ri, is a function of the ambient air Richardson number, $Ri_a$, and a term involving the stratification of the cloud.

The assumptions regarding the vertical dispersion coefficient, $K^m_z$, are arbitrary but reasonable, and probably satisfy the requirement that the effective scale of the turbulence be less than the scale of the cloud. But the form of the equation for the horizontal dispersion coefficient, $K^m_y$, would lead to much discussion among modelers. Many modelers prefer to use a $K^m_y$ formulation such that $K^m_y$ is independent of height. Intuition would lead to the conclusion that $K^m_y$ would be larger for very dense clouds, while equations (5-45), (5-46), and (5-49) would seem to lead to the opposite conclusion.

An advantage of numerical models is that they can more precisely treat sources which have time and space variability in their emission rate. For example, the emissions from individual grids covering an evaporating pool can be given as a function of time. Other models of this type have been proposed by Deaves (1984), England et al. (1978), and Riou and Saab (1985), and will be summarized in Section 5.2. These models can produce voluminous output information, as illustrated by the wind field predictions for a Thorney Island trial by Riou and Saab (1985) in Figure 5-8. A drawback, of course, is the need for a mainframe computer and the errors introduced by numerical approximations.

## Physical Models

Because of the difficulties in mathematical modeling and conducting full-scale field experiments on transport and dispersion of hazardous materials, much research has been done with physical models, i.e., wind tunnel or water channel laboratory simulations. Work in this area has been reviewed by Meroney (1982) and Duijm et al. (1985). These studies have emphasized dense gas dispersion from continuous elevated sources and from instantaneous ground level sources. A major concern is whether geometrical and physical parameters have been properly scaled. It is found that some of these scaling requirements may be relaxed somewhat; for example, if the Reynolds number is higher than about 1000, then the flow is fully turbulent. A fundamental requirement in dense gas studies is that the densimetric Froude number must be simulated:

$$Fr = u/(gl(\rho_p - \rho_a)/\rho_a)^{1/2} \qquad (5\text{-}50)$$

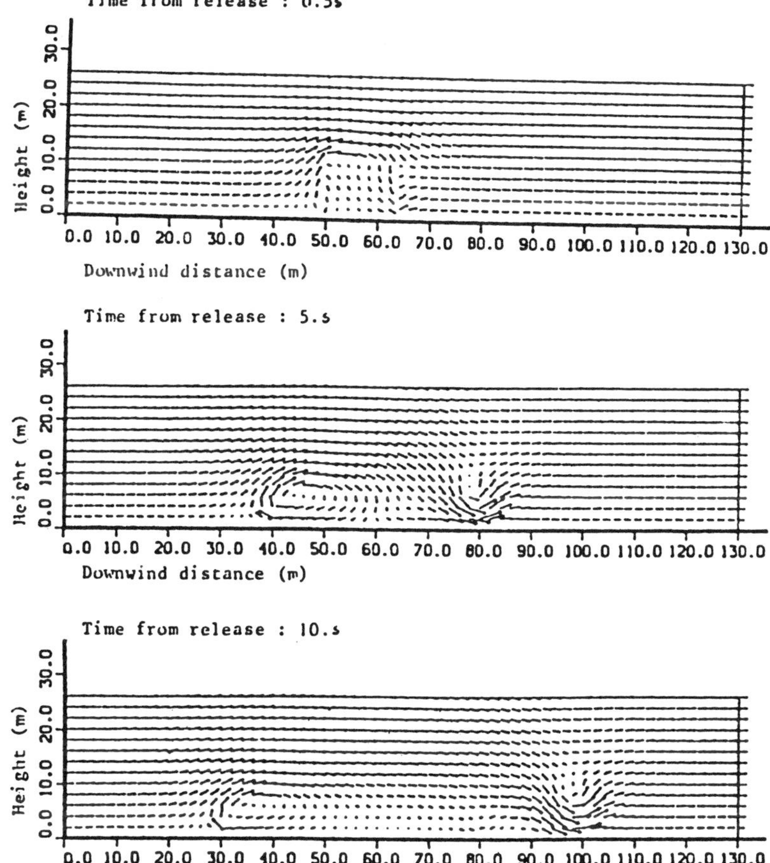
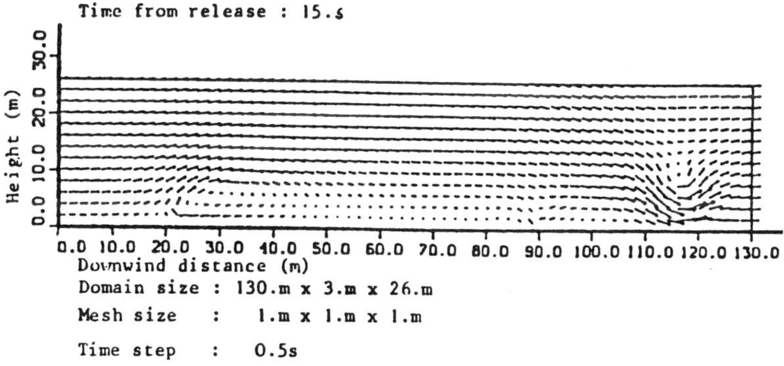



Figure 5-8. Calculated wind vector field (Mercure-Gaz Lourds) from the numerical model by Riou and Saab (1985). The arrows represent wind vectors at grid points. The source is a square with side 10m.

where l is the length scale of the source and u is the wind speed. Note that Fr is related in an inverse way to the source Richardson number, $Ri_o$, defined earlier in equation (5-13). Very close to the source, simulation of the relative density difference, $(\rho_p - \rho_a)/\rho_a$ is also important. Once the scaling laws are properly followed, the dispersion in the physical model can closely simulate dispersion in the field, as illustrated in Figure 5-9 from Meroney (1982).

Havens and Spicer (1985) used physical modeling to help develop and test their DEGADIS model for dense gas dispersion. The most effective use of physical modeling is to study special situations where field studies are impractical, such as the dispersion in the vicinity of a complex of buildings, a dike, or at a complex terrain site. Meroney (1979) has used his wind tunnel to study the special question of the "lift-off" of buoyant plumes released at the ground and transported in a shear flow. He concludes that the distance to lift off is given by the expression:

$$x \simeq 0.24[g((\rho_p - \rho_a)/\rho_a)V'_o/u^3]^{-0.5} \; R_o^{1.5} \tag{5-51}$$

where $V'_o$ is the initial plume continuous volume flux and $R_o$ is the initial source dimension. This lift-off distance increases as wind speed increases, and decreases as the cloud buoyancy increases, as expected from intuition.

5.1.5 Non-Homogeneous Terrain and Wind Fields

Most of the hazardous gas models described above are applicable only to flat unobstructed terrain, steady state homogeneous wind fields, and constant ambient stability. Unfortunately, accidental releases of hazardous gases do not restrict themselves to these simplified conditions. In particular, dense gas clouds will flow towards the slightest depression in the terrain, and dense gases have been observed to flow upstream, against the wind, if the terrain has sufficient slope in that direction. The models available for these non-homogeneous conditions can be divided into four basic groups.

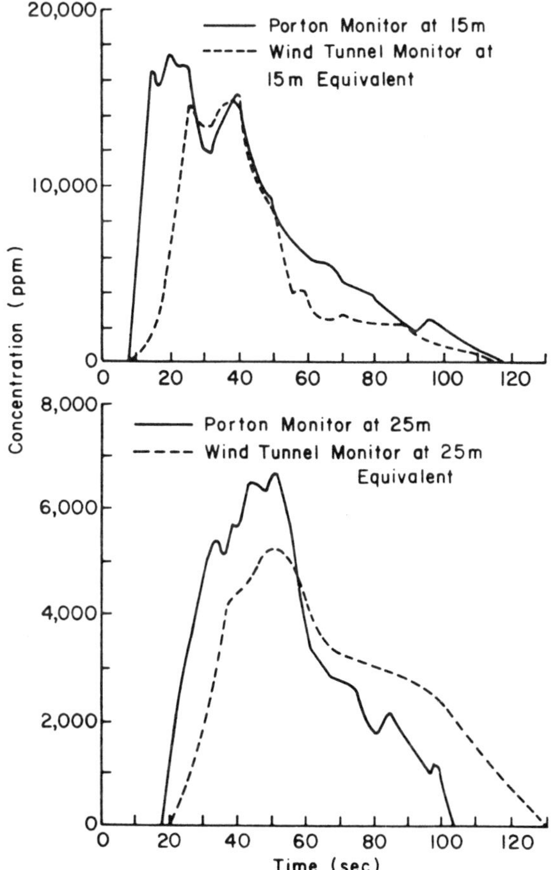



Figure 5-9. Comparison of field data (Porton) and wind tunnel simulations of dense gas dispersion (from Meroney 1982).

- Standard EPA Gaussian models that are corrected for plume lifting as the plume approaches a hill.

- Drainage flow models (for shallow dense gas flow down slopes).

- Three dimensional models.

- Puff trajectory models (for non-homogeneous wind fields only).

The first category of models was developed by the EPA for treating situations where the elevation of nearby hills is greater than the plume elevation. These models only apply to plumes which have become neutrally-buoyant by the time they reach the terrain. Figure 5-10 shows the plume lifting assumptions for stable conditions and unstable/neutral conditions in the standard EPA COMPLEX I and II models (available from NTIS). No lifting is assumed for stable conditions, although the model does not permit the plume centerline to approach within 10m of the ground. For unstable/neutral conditions, the model assumes that the plume is lifted so that its initial height above the terrain, $z_e$, is reduced by $z_e/2$ or $z_h/2$, whichever is less (where $z_h$ is the hill height). More advanced models are being developed that account for the effects of atmospheric stability on plume lifting through the calculation of a critical height, $h_c$:

$$h_c = z_h \ (1.0 - u/((g/T)(dT/dz + 0.01 \ ^oK/m)))^{1/2})$$
(5-52)

If the plume height, $z_e$, is below this critical height, $h_c$, then impaction is assumed (as in Figure 5-10a), and if it is above $h_c$, than lifting is assumed (as in Figure 5-10b). However, the lifting in this case would be scaled to the difference $z_e - h_c$ rather than $z_e$. The new CTDM model being developed by the EPA will further improve these categories of models by including analytical solutions for the streamline deflection around the hill (Strimaitis et al., 1986).

Drainage flow models are concerned with the nocturnal cooling of slopes and the subsequent formation of shallow gravity-driven density flows. This

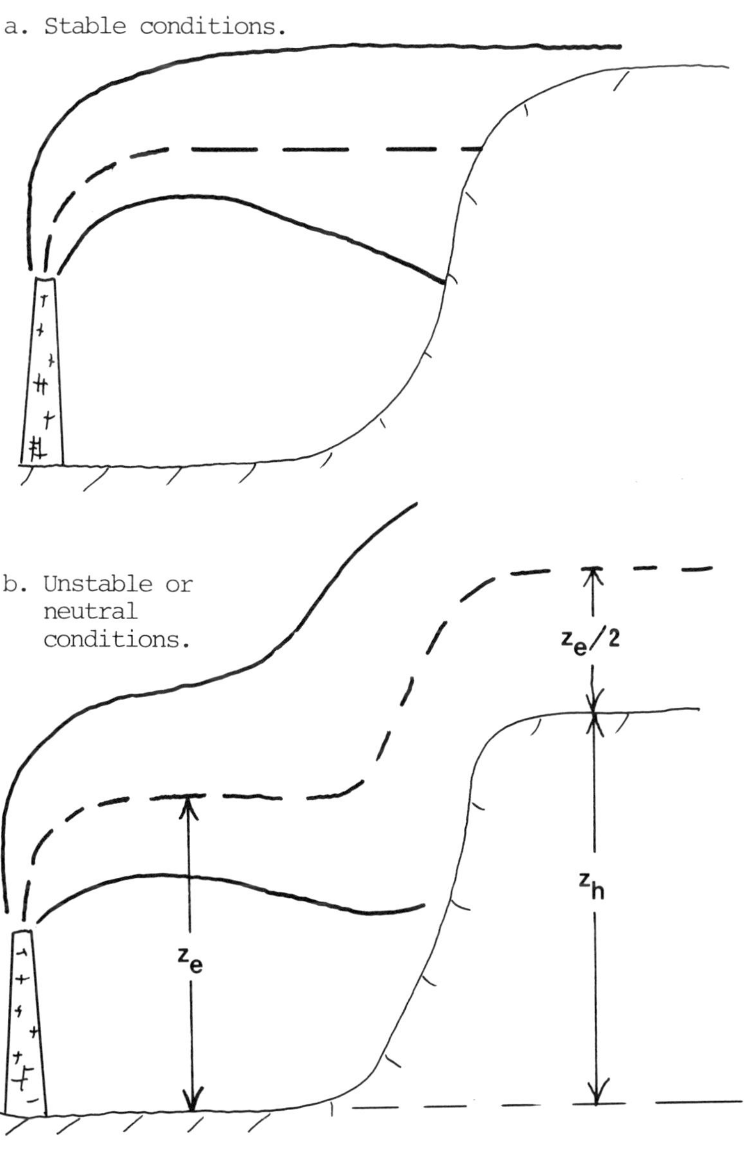

Figure 5-10.   Plume lifting assumptions in EPA models for complex terrain.
(The dashed line is the plume centerline.)

situation is analogous to dense gas flow down slopes and the models could probably be modified to account for dense gas sources. Horst and Doran (1986) analyze the energy budget of a drainage flow, showing how a shallow layer is developed containing a relatively strong wind jet and large turbulent energy. These flows are typically only a few meters deep for downslope distances from the hilltop of about 100m, and can reach depths of about 100m for downslope distances from the hilltop of several kilometers. DeNevers (1984) applied a simple energy balance formula to the case of a dense gas cloud moving down a slope, and showed that the velocity due to downslope effects would equal the frontal velocity due to dense gas slumping for slope angles of only a few degrees. A limitation of the models is that they generally assume lateral homogeneity, whereas real life slope flows usually are influenced by geographical features such as valley walls.

Most three dimensional models can handle inhomogeneities, energy transfer processes and roughness effects associated with the underlying terrain. In fact, this is one of the most rapidly growing research areas in micro- and meso-meteorology, and has been extensively discussed in Pielke's (1981) review of mesoscale modeling. In general the term micro applies to distance scales less than about 10km, and the term meso applies to scales from about 10 to 1000 km. Yamada's (1985) three-dimensional model is currently being implemented on microcomputers at Department of Energy sites, although it does not explicitly treat dense gases. A terrain-following coordinate system in the vertical direction is used to simplify the treatment of surface boundary conditions, and dispersion is calculated using a random-particle statistical model. Yamada's (1985) predictions of a the shape of a non-buoyant plume at a specific time over the Brush Creek Valley, Colorado, area are shown in Figure 5-11. These results look fairly plausible, showing increased dispersion at mid-day.

The three dimensional dense gas models by Chan and Ermak (1985) and Riou and Saab (1985) are capable of treating complex terrain. But even though "complex or variable terrain" appears in the titles of their articles, their texts contain little information on how the models handle the variable terrain. Chan and Ermak (1985) give examples of the predictions of their FEM3 model with and without terrain for the Burro 8 field test, showing slight

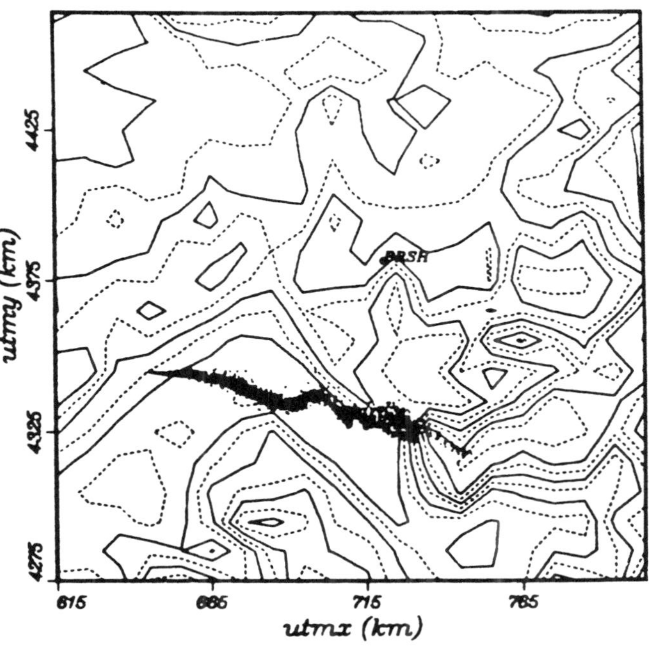


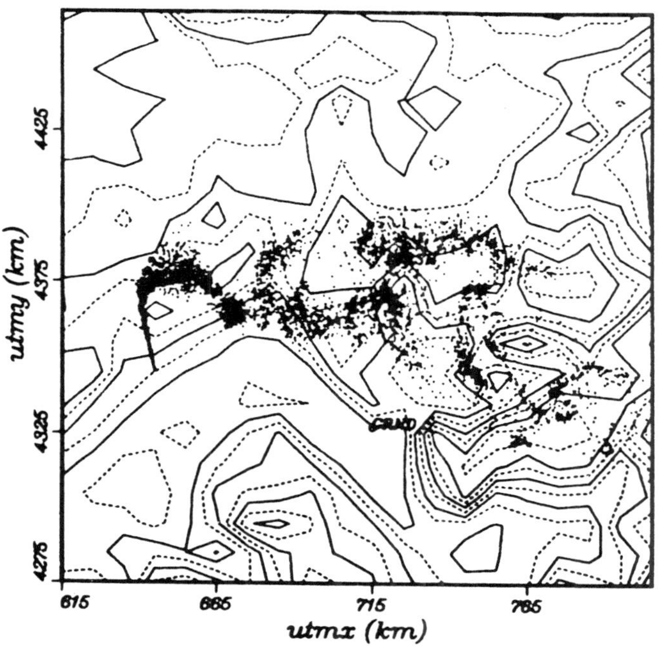


Figure 5-11.  Simulated plume dispersion for 1000 hr (left diagram) and 1400 hr (right diagram) at Brush Creek Valley, CO, from Yamada (1985). Solid terrain contours are shown at 400m intervals.

variations in predicted concentrations. For example, the observed distance to the lower flammability limit was 440m while the predictions of the flat terrain and variable terrain simulations were 470m and 400m, respectively. At that site, however, the terrain showed only minor variations in height.

The three-dimensional model, HEAVYGAS, by Deaves (1985) can account for flow around obstacles such as buildings and fences. It was applied to the Thorney Island Phase II trials in which a 10m cube "building" was placed 20m upwind of the dense gas release. The program accounts for the development of the building wake and the perturbation in turbulence intensity.

Space and time variations in the wind field can occur even over flat terrain and are caused by mesoscale eddies generated by synoptic disturbances, uneven surface radiation effects, or internal waves. A typical root mean square difference in wind direction between stations in a mesoscale network is about $30^{\circ}$ (Hanna et al 1982). In complex terrain, mountain-valley flows and channeling can lead to even greater differences. Observations from the Idaho National Engineering Laboratory (INEL) are plotted in Figure 5-12, from a report by Lewellen et al. (1985), showing the great spatial and hour-to-hour variability in the wind field at that site. The observed tetroon (constant level balloon) trajectories plotted on the figure also indicate this variability, executing loops and showing that different trajectories from the same morning can be $90^{\circ}$ to $180^{\circ}$ from each other. The variability of winds at this site is of great practical importance, since hazardous gases (radioactive) are routinely released on the site and the potential exists for accidental releases of larger quantities of hazardous materials.

The observed constant-level balloon trajectories in Figure 5-12 point out that emergency planning may not be reliable if only a single wind observation is made, and can often be completely unreliable if the wind observation site is located 10km or more from the hazardous gas release point. As an alternative to a full three-dimensional model, many researchers are using puff models based on wind fields derived from interpolated wind observations. The wind fields are handled differently by the different models. For example, the EPA model

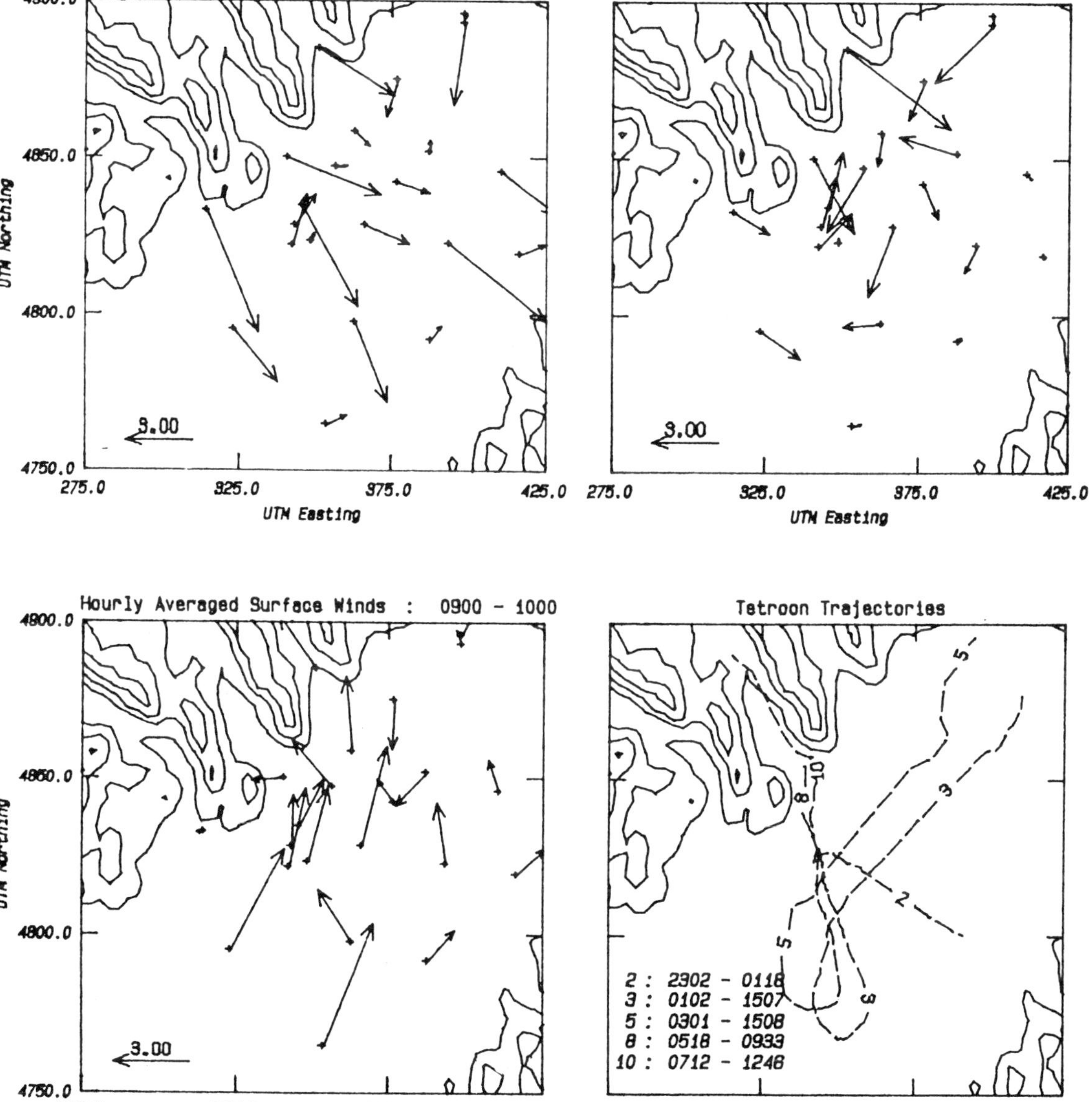


Figure 5-12.  Wind data from the Idaho National Engineering Laboratory
(Lewellen et al 1985).  The scale (in m/s) is in the lower
left corner of the first three diagrams.  Observed tetroon
(constant level balloon) trajectories are in the last
diagram.  The arrows labelled "3.00" refer to a speed scale
of 3.00 m/s.  Contour interval is 500m and horizontal axes
are given in km.

INPUFF 2.0 (Petersen and Lavdas, 1986) simply assumes that the wind field is "input by the user". At the other extreme, Ludwig's (1985) model generates a three dimensional wind field using available data and mass-continuity principles. A relaxation procedure is used to obtain the nondivergent wind field. Midway between these extremes is the procedure used in the EPA model MESOPUFF II (Scire and Lurmann 1983), which interpolates the wind speed components $(u_s, v_s)$ at any point from the formula

$$(u_s,v_s) = \sum_k (\alpha_s/r_s^2) \cdot (u_k,v_k) / \sum_k (\alpha_s/r_s^2) \tag{5-53}$$

where $(u_k,v_k)$ are the easterly and northerly components of the wind at monitoring site $k$, $r_s$ is the distance from the point of interest to the monitoring site, and $\alpha_s$ is an alignment weighting factor ($\alpha_s = 1-0.5|\sin\phi_s|$), where $\phi_s$ is the angle between the observed wind direction and the line from the monitoring site to the point of interest.

Once the wind field is known, then the trajectories and dispersion of puffs or plume segments can be calculated. It is assumed that the mass emitted in a short period (say 15 minutes) is transported and dispersed as a single puff. The center of that puff moves with the wind at that particular location and time. If the winds are variable, the puffs emitted in subsequent time periods may travel in different directions, as illustrated in Figure 5-13, from the INPUFF-2.0 Users Guide by Petersen and Lavdas (1986). If it is assumed that lateral and longitudinal dispersion are equal, then the ground level concentration in each puff can be written as

$$C = (Q_i/((\sqrt{2}\pi^{3/2}\sigma_r^2\sigma_z)) \exp(-0.5r^2/\sigma_r^2) \exp(-0.5z_e^2/\sigma_z^2) \tag{5-54}$$
$$\text{for } \sigma_z < 0.8h$$

$$C = (Q_i/(2\pi\sigma_r^2 h)) \exp(-0.5r^2/\sigma_r^2) \text{ for } \sigma_z > 0.8h \tag{5-55}$$

where $r^2 = (x-ut)^2 + y^2$, $\sigma_r^2 = \sigma_y^2 + \sigma_x^2$, and $Q_i$ is the mass of pollutant emitted in the time interval. Puff models have special procedures to decide the optimum time interval for emission of each puff and to combine puffs after

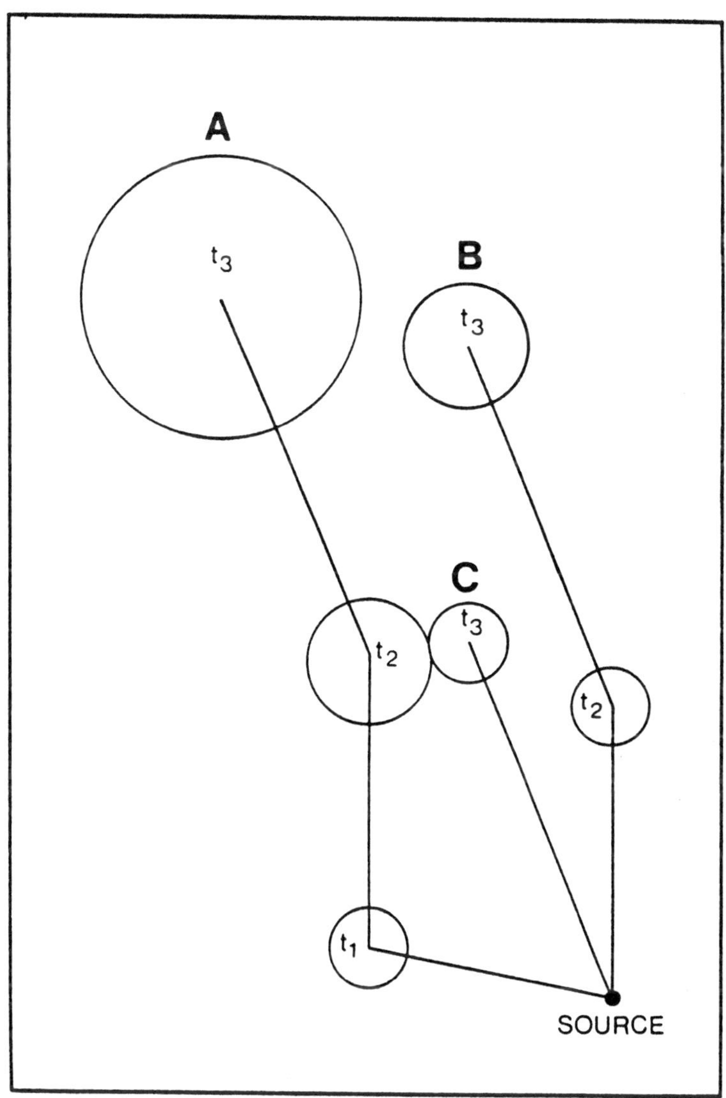



Figure 5-13.  Hypothetical puff trajectories, where puffs A, B, and C are emitted at the beginning of time periods 1, 2, and 3, respectively (from Petersen and Lavdas, 1986).

a sufficient time has elapsed that neighboring puffs overlap each other. This computational problem has been partially solved by Zannetti (1986), who uses trapezoidal-shaped plume segments rather than puffs under certain conditions in his AVACTI model.

While the puff trajectory models are justified from physical arguments, Lewellen et al. (1985) have found that their predictions do not agree with observations much better than the predictions of standard straight-line Gaussian plume models. They used data from the set of experiments at Idaho National Engineering Laboratory (INEL) from which the example in Figure 5-12 was drawn. It is suspected that the uncertainties due to stochastic variability in atmospheric concentrations are responsible for the lack of a significant difference among the models.

5.1.6 Concentration Fluctuations; Averaging and Sampling Time

Discussions of averaging and sampling times (see Figure 2-1) or concentration fluctuations are either non-existent or very minimal in the research reviewed to this point. Some models grossly parameterize this effect by assuming that the ratio of the peak (fluctuating) concentration to the model predicted mean concentration is about two. Chatwin (1982) pointed out that in many cases involving accidental releases of hazardous gases, the maximum short term (~1 sec) concentration is the most important variable to predict. Lung damage from $H_2S$ can occur with one breath if the concentration is sufficiently high, and an explosion of gas from an LNG accident can occur if a spark is struck in a small volume of gas at the flammability limit. According to Chatwin the mean concentration predicted by the model can be irrelevant in these cases, since the probability distribution function (pdf) of concentration fluctuations in the atmosphere is characterized by a standard deviation at least as large as the mean. The relative magnitude of concentration fluctuations ($\sigma_c/\bar{C}$) is the same order as the relative magnitude of velocity fluctuations ($\sigma_u/U$) in the atmosphere. The parameters $\sigma_c$ and $\sigma_u$ are the standard deviations of turbulent fluctuations in concentration and wind speed, respectively. Thus it is important to predict the upper end of the pdf for the

$H_2S$ and LNG incidents described above. Since Chatwin's article was published, a few other researchers have studied this problem, although we are far from having a comprehensive operational model.

Predictions of models such as DEGADIS or FEM3 can be thought of as ensemble means for certain averaging times. An ensemble mean is defined as the mean over an infinite number of realizations of a given experiment. The averaging time is usually implicit in the data used by the model and in its formulations for treating the input data - for example, if hourly-averaged wind and turbulence observations are used, then the predictions represent a one hour average. If the Pasquill-Gifford-Turner dispersion curves are used, then the predictions represent a 10 minute average, since data from 10 minute periods were used to derive the curves. In the case of instantaneous (puff) models, the predictions represent an ensemble mean only to the extent that a large enough set of experiments (20 or more) was used to derive the model. These experiments should be conducted under the same external conditions (i.e., wind speed, stability, source term). For example, if it were possible to run the Thorney Island experiments long enough that 100 independent time periods (e.g., of ten-minute duration) could be found which all satisfy the following conditions:

$$4.8 < u < 5.2 \text{ m/s}, \qquad\qquad 65\% < RH < 70\%$$
$$10^{O} < T_a < 12^{O}C, \qquad\qquad 10^{O} < T_{\text{surface}} < 12^{O}C$$
$$-2 < \text{net radiation flux} < 2 \text{ watts/m}^2$$
$$(\rho_p - \rho_a)/\rho_a = 2, \quad h=10m, \quad R=10m,$$

then the observed concentration field averaged over these 100 experiments would approach an ensemble average. The reader quickly sees that it is difficult operationally and financially to generate ensemble averages from atmospheric field experiments. A true ensemble would contain an infinite number of individual data points!

Thus the results of a single experiment, or even three or four experiments will likely differ (perhaps by as much as an order of magnitude) from the ensemble mean predictions of the model. If this happens, it is not an indictment of the model but may be a manifestation of the inherent stochastic variability of the atmosphere.

Wind tunnel experiments can be used to study variability, since it is easier to insure repeatability of experiments and thus create a large ensemble of data. On the negative side, the wind tunnel cannot simulate larger scale eddies and other phenomena that contribute to variability in the atmosphere. Furthermore, the laboratory Reynolds number is not high enough to permit the establishment of an inertial subrange like there is in the atmosphere. Meroney and Lohmeyer (1984) conducted extensive studies of dense gas clouds released in a wind tunnel and calculated the concentration fluctuation intensity, $\sigma_c/\overline{C}$, for various source volumes, wind speeds and downwind distances. These results are plotted in Figure 5-14, showing that the average $\sigma_c/\overline{C}$ is about 0.3 in this wind tunnel. In contrast, Hanna (1984) reports observed values of $\sigma_c/\overline{C}$ of 1.5 on the plume centerline and $\sigma_c/\overline{C}$ of 5.0 on the plume edges for a smoke plume released in the atmospheric boundary layer.

The probability distribution function (pdf) of concentration fluctuations in the atmosphere has been studied by several persons (e.g. Wilson, 1982; Hanna, 1984; Lewellen and Sykes, 1985), and all agree that the distribution is non-Gaussian and is skewed towards higher concentrations. For hazardous gas analysis, we are usually interested in the probability $P(C>C_L)$ that the concentration is higher than some limiting value, $C_L$:

$$P(C>C_L) = \int_{C_L}^{\infty} p(C)\ dC \tag{5-56}$$

It has been suggested that the probability distribution function, $p(C)$, can be approximated by a log-normal, clipped normal, or Gamma function. The exponential function is a special case of the Gamma function, and is quite good for intermittent clouds or plumes. The intermittency, I, is defined as the fraction of time that non-zero concentrations are observed at a monitor. For the exponential distribution, $\sigma_c/\overline{C}$ equals one. In this case, the pdf is given by the formula:

$$p(C) = (I^2/\overline{C})\exp(-IC/\overline{C}) + (1-I)\,\delta(0) \tag{5-57}$$

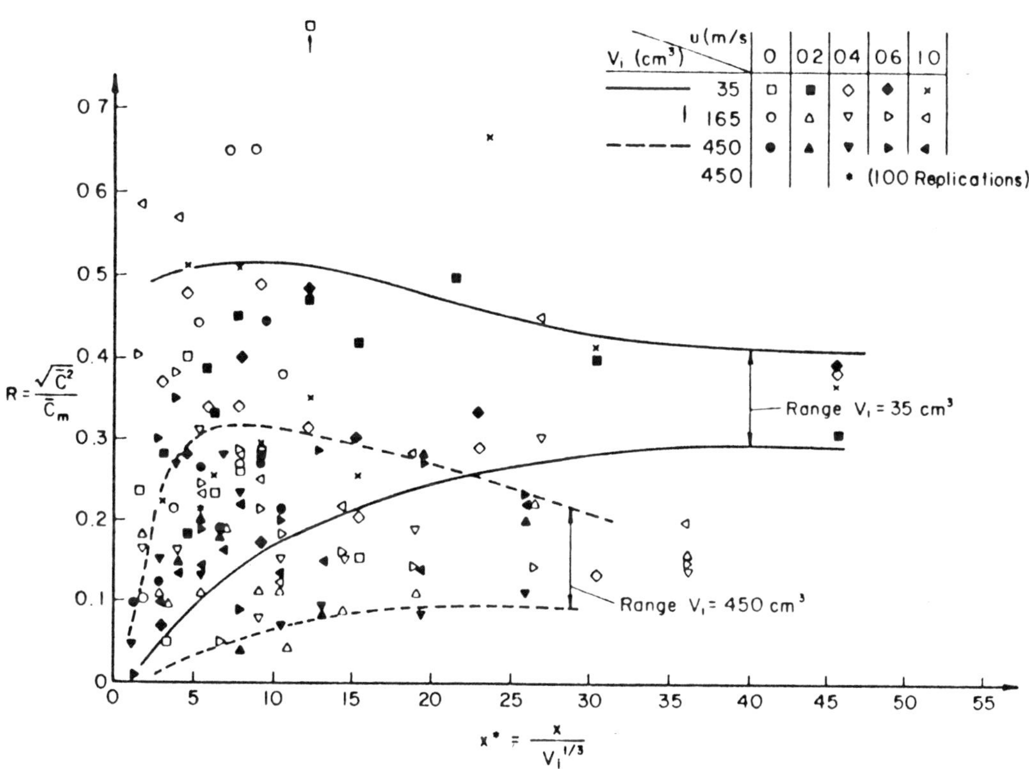


Figure 5-14.  Concentration variance ratio, $\sigma_c/\bar{C}$, versus downwind distance, observed by Meroney and Lohmeyer (1984) in a wind tunnel. The source is an instantaneous dense gas cloud.

where the Dirac delta function $\delta(\emptyset)$ equals 1.0 at C equal to $\emptyset$ and equals 0.0 elsewhere. This can be substituted into equation (5-56) to give:

$$P(C>C_L) = I \exp(-IC_L/C) \tag{5-58}$$

Thus if the dispersion model predicts an ensemble mean, $\overline{C}$, of $0.1C_L$ , where $C_L$ is the threshold concentration for some health effect, and the intermittency I equals 0.5, then the probability that the instantaneous C will exceed $C_L$ is 0.3%. If the ensemble mean prediction is $0.5C_L$, then this probability is 18%.

The formulas given above are for nearly-instantaneous averaging times. It is clear that the standard deviation of concentration fluctuations, $\sigma_c$, will decrease as averaging time T increases. If the integral time scale of the concentration fluctuations is $T_I$ and the autocorrelogram is assumed to be exponential, then the following formulas apply:

$$R(t') = \overline{C'(t)C'(t+t')}/\sigma_c^2 = \exp(-t'/T_I) \tag{5-59}$$

Then

$$\sigma_c^2(T)/\sigma_c^2(\emptyset) = 2(T_I/T)(1-(T_I/T)(1-\exp(-T/T_I))) \tag{5-60}$$

where $\sigma_c^2(0)$ refers to the variance for instantaneous averaging time. If $T_I$ has a typical value for the surface layer (about 10 sec), then the ratio of variances for an averaging time of T equal to 60 sec is 0.28. If the averaging time is one hour, the ratio $\sigma_c(3600s)/\sigma_c(0)$ is 0.075. It can be concluded that the fluctuation intensity $\sigma_c/\overline{C}$ for one hour averages in the atmosphere is about 0.1 even if the integral time scale is only a few seconds.

If it is assumed that the equations in the first part of this section produce predictions of ensemble mean concentrations, $\overline{C}$, then the probability of the concentration exceeding any threshold limit, $C_L$, can be estimated using equations (5-56) through (5-60) for any averaging time and integral time scale.

Equation (5-60) can be used to assess the effects of averaging over distances as well as time. Observed concentrations and health effects always involve some averaging distance. For example, if the integral distance scale of the turbulence is 5m and the averaging distance is 1m, then the ratio $\sigma_c^2(1m)/\sigma_c^2(0)$ equals 0.94.

At the other end of the scale the sampling time or sampling volume can also influence observations. Figure 2-2 illustrates a typical time series, showing that the sampling time, $T_s$, can be thought of as the total length of time that the instrument is turned on. It is intuitively obvious that the likelihood of more extreme concentrations being observed is increased if the sampling time increases (e.g. notice how several new "record" high and low temperatures are observed at any given weather station each year). The usual definition of any ensemble assumes that the sampling time is infinity. In practice this requirement is considerably relaxed, such that a set of ten dense gas experiments conducted during similar external conditions is assumed to comprise an ensemble. Equation (5-60) can also be used to calculate the variance "missed" by an instrument because it is turned on for a finite sampling time $T_s$:

$$\sigma_c^2(0, \ T_s)/\sigma_c^2(0, \infty) = 1-2(T_I/T_s)(1-(T_I/T_s)((1-\exp(-T_s/T_I)))) \tag{5-61}$$

where the first variable inside the parentheses after $\sigma_c^2$ is the averaging time and the second variable is the sampling time. Any eddies with time scales much larger than $T_I$ are not detected by the instrument. For example, if $T_s$ is ten times the integral scale $T_I$, then only 82% of the total possible variance is seen. If both the sampling time $T_s$ and the averaging time T are finite (as they are in any experiment) then the fraction of the total possible variance can be calculated by multiplying equations (5-60) and (5-61) together. An example is given in Figure 5-15 for the special case $T_s/T = 100$ (for example, averaging time could be one minute and sampling time could be 100 minutes). This function clearly defines a "window", with high and low frequency fluctuations filtered out by the finite sampling and averaging times.

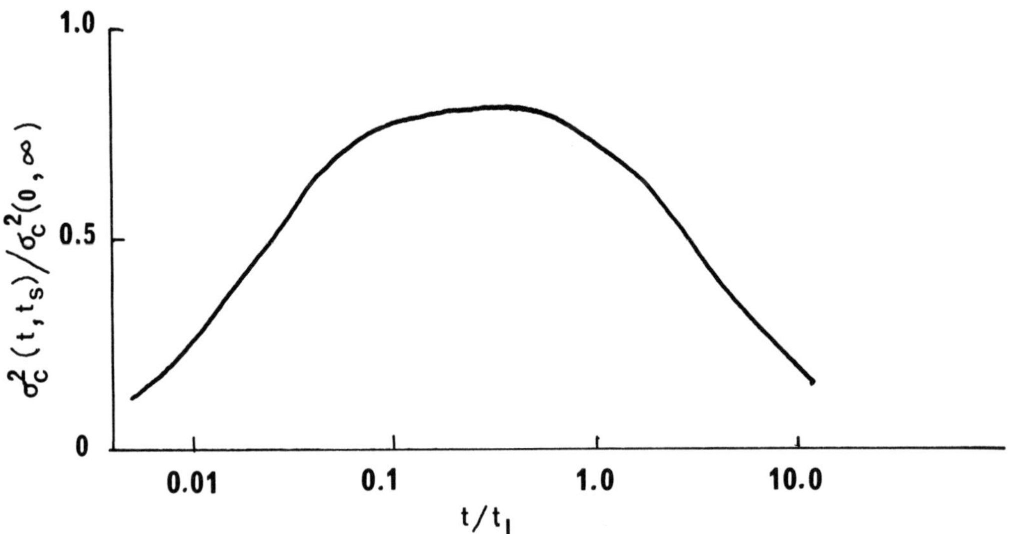


Figure 5-15. Fraction of total possible concentration fluctuation variance from Eqs (5-60) and (5-61), as a function of sampling time $T_s$, averaging time $T$ and integral time scale $T_I$. It is assumed that $T_s/T_I = 100$.

## 5.2 Review of Procedures used by Specific Models for Transport and Dispersion

The previous subsection discussed specific models only insofar as they were useful to illustrate basic physical principles important for the transport and dispersion of hazardous materials. Generally for every model used as an illustration, there are many more similar models available in the literature. This subsection contains two components: a discussion of four specific models, and a table summarizing the characteristics of similar models. The reader is reminded that this book represents a snapshot of the models as of Fall 1986, and most models are in a continual state of modification. Furthermore, omission of some models is due to the failure of the authors to obtain literature on the models, for which we apologize. Finally, if several models are very similar, only one will be presented in the table.

### 5.2.1 Four Models for Hazardous Gases

The four models that are discussed below are two slab or similarity models for dense gases (AIRTOX and DEGADIS), a three-dimensional model for dense gases (FEM3), and a puff trajectory model for neutrally-buoyant gases (INPUFF 2.0). None of these models handles two phase jets or clouds or multicomponent spills.

The choice of these four models for discussion does not imply any endorsement of the models. The main criterion for inclusion here is that the published reports on the models contain a reasonably complete description of their governing equations. There are many other excellent models which are not discussed in this section but which are summarized in Appendix A.

### AIRTOX

The AIRTOX model for the release and dispersion of toxic air contaminants is described by Paine et al. (1986). It can account for single-phase denser-than-air releases and uses the box or slab model approach to calculate dispersion, incorporating various theoretical models developed by a number of researchers.

The model includes predictions of the momentum jet from a high-velocity gas release. The radius of the jet equals $0.16x$ and the length of the jet stage is $(V'_o w_o)^{1/2}(0.16u)^{-1}$, where $V'_o$ is the initial volume flux (divided by $\pi$) and $w_o$ is the initial velocity. The liftoff of a buoyant plume attached to the ground is given by equation (5-51). If the plume is dense and is released from an elevated stack at height $z_e$, the distance, $x_g$, to the point of touch down is assumed to equal:

$$x_g = (z_e^3 u^3 / (V'_o g (\rho_p - \rho_a) / \rho_p)) \tag{5-62}$$

<u>Instantaneous source</u>: The pseudo-instantaneous dense gas cloud is assumed to have an initial height equal to 0.5 times the radius, and the spread velocity is given by equation (5-16), which appears to be universally applied in slab models $(dR/dt \propto (g((\rho_p - \rho_a)/\rho_a)V^{1/3})^{1/2})$. Transition to passive dispersion occurs at the distance where the radial velocity equals the lateral turbulent velocity standard deviation $(\sigma_v)$:

$$dR/dt = \max\ (2u_*,\ 0.5m/s) \tag{5-63}$$

All box or slab models require an entrainment assumption, and AIRTOX uses the Zeman (1982) approximation, $dV/dt = \pi R^2 w_e$, where

$$w_e = 3.5\ v_*/(11.67 + Ri) \tag{5-64}$$

such that entrainment into the top of the slab is reduced as the local Richardson number,

$$Ri = (g(\rho_p - \rho_a)/\rho_a)h/v_*^2 \tag{5-65}$$

increases. The characteristic vertical turbulent velocity, $v_*$, is given by the formula:

$$v_* = 1.3(u_*/u)((4/9)(dR/dt)^2 + u^2)^{1/2} \tag{5-66}$$

Note that this contains a component due to ambient turbulence and a component due to the spread of the cloud.

<u>Continuous source</u>:  If the dense gas source is continuous, then the treatment suggested by Raj (1985) is used.  The initial plume width is assumed to be given by the equation:

$$R_{yo} = g((\rho_p - \rho_a)/\rho_a) V_o'/2u^3 \qquad (5\text{-}67)$$

where $V_o'$ is the initial volume flux in $m^3/s$.  The plume width then grows with distance according to the formula:

$$R_y = R_{yo}(1+2.1(x/R_o u)(g((\rho_p - \rho_a)/\rho_a)h_o)^{1/2})^{2/3}) \qquad (5\text{-}68)$$

Volume flux equals $2R_y hu$ and the variation of the plume height h with time is given by $dh/dx = w_e/u$, where $w_e$ is given by equation (5-64).  Thus the entrainment rate is parameterized in the same way for instantaneous and continuous sources.

The passive plume is assumed to disperse according to standard Gaussian formulas for puffs and plumes, where the dispersion parameters at the transition point are given by the formulas:

$$\sigma_{xo} = \sigma_{yo} = R_y/\sqrt{2}; \quad \sigma_{zo} = h/\sqrt{\pi/2} \text{ for puff}$$

$$\sigma_{yo} = 2R_y/\sqrt{2\pi}; \quad \sigma_{zo} = h/\sqrt{\pi/2} \text{ for plume}$$

Virtual source distances are calculated for both $\sigma_{yo}$ and $\sigma_{zo}$.  Subsequent dispersion of the puff in the x direction also contains a wind shear component following the recommendations of Wilson (1981), since the dense gas puff is located in the surface layer where the wind shear acts to stretch out the puff:

$$\sigma_x^2 = \sigma_x^2 \text{ (Pasquill-Gifford)} + 0.09 \, \sigma_z^2 (nxz_r^{n-1}z_c^{-n})^2 \qquad (5\text{-}69)$$

where n is the power law exponent for the wind profile ($u \propto z^n$) and $z_r$ and $z_c$ are reference heights fully described by Wilson (1981).

The AIRTOX model also contains algorithms for estimating some types of source emissions.  Paine et al. (1986) and Heinold et al. (1986) show that the model compares fairly well with atmospheric field data.  It is a true hybrid model, being composed entirely of pieces brought together from a wide variety of independent studies.

<u>DEGADIS</u>

The DEGADIS (DEnse GAs DISpersion) model was developed by Havens and Spicer (1985) and is an adaptation of the Shell HEGADAS model described by Colenbrander (1980) and Colenbrander and Puttock (1983), and a theoretical paper by van Ulden (1983).  It assumes that the cross-wind distribution of material is uniform in the middle of the cloud and is Gaussian at the edges, and the vertical distribution of material follows a modified power law. The $\sigma_y$ and $\sigma_z$ coefficients are not the same as those used in EPA models.  This model is relatively simple in concept but exhaustive in detail.  The highlights of the model are covered below, and details on all components are given in the references.

The dense gas cloud near the source is represented as a cylindrical gas volume which spreads laterally as a density-driven flow with entrainment at the top of the cloud by wind shear and air entrainment into the advancing front edge.  The model does not handle releases with any excess momentum.  The frontal (spreading) velocity is modeled as

$$U_f = C_E(g(\rho - \rho_a)h/\rho_a)^{1/2} \qquad\qquad (5\text{-}70)$$

where the value of $C_E$ used is based on laboratory measurements of cloud spreading velocity.  The radius of the cloud is constrained to be greater than or equal to the radius of any primary (liquid) source present.

The DEGADIS model has a special algorithm for calculating the initial dense cloud behavior.  This algorithm accounts for the mass balance right over the evaporating liquid and can handle slowly or rapidly boiling liquids.  The term "source blanket" is used, which refers to an accumulation of mass over a rapidly boiling liquid.  The enthalpy of the emitted gas, the enthalpy of the ambient humid air, and the enthalpy of any water vapor entrained by the blanket are included.  There are three alternate submodels included for the heat transfer from the surface to the cloud.  The standard approach uses "handbook" correlations to estimate the heat transfer coefficient.

If a steady state spill is being simulated, the transient source calculation is carried out until the source characteristics are no longer varying significantly with time. The maximum centerline concentration $C_o$, the horizontal and vertical dispersion parameters $\sigma_y$ and $\sigma_z$, the half width R, and if necessary, the enthalpy h are used as initial conditions for the downwind calculation specified in a transient spill.

If a transient spill is being simulated, it is modeled as a series of pseudo-steady state releases. Consider a coordinate system traveling with the wind over the transient gas source described above and originating from the point which corresponds with the maximum upwind extent of the gas blanket. A series of plumes are calculated for several time segments that make up the total time period of interest and the results are added to give the mean prediction. This approximation should be evaluated by comparisons with true transient models.

For steady-state conditions, the vertical dispersion parameter $\sigma_z$ is determined by requiring that it satisfy the diffusion equation with the vertical turbulent diffusivity given by

$$K_z = 0.4 \; u_* z / \phi (Ri_b) \tag{5-71}$$

The function $\phi (Ri_b)$ is a curve fit of laboratory scale data for vertical mixing in stably-stratified fluid flows reported in the literature for $Ri_b > 0$. For $Ri_b < 0$, the function $\phi (Ri_b)$ is taken from Colenbrander and Puttock (1983) and has been modified so the passive limit of the two functions agree as follows:

$$
\begin{aligned}
\phi (Ri_b) &= 0.88 + 0.099 \; Ri_b^{1.04} + 1.4 \times 10^{-25} \; Ri_b^{5.7} \qquad & Ri_b \geq 0 \\
&= 0.88/(1 + 0.65 \; Ri_b^{\;0.6}) & Ri_b \leq 0
\end{aligned}
\tag{5-72}
$$

where the bulk Richardson number $Ri_b$ is computed as

$$Ri_b = g(\rho - \rho_a) h_e / (\rho_a u_*^2) \tag{5-73}$$

and the effective cloud depth, $h_e$, is defined as

$$h_e = C_c^{-1} \int^{\infty} C dz = \Gamma((1+\alpha)^{-1}) \sigma_z / (1+\alpha) \qquad (5\text{-}74)$$

where $C_c$ is the centerline concentration and $\alpha$ is the power law appropriate for the vertical concentration distribution. The vertical $\sigma_z$ is calculated using equation (6) of Colenbrander (1980). After some manipulations, it can be shown that

$$d(\rho u_e h_e)/dx = 0.4 \, \rho_a u_* (1+\alpha)/\phi(Ri_b) \qquad (5\text{-}75)$$

for two-dimensional dispersion. When heat transfer from the surface is present, vertical mixing will be enhanced by the convective turbulence due to heat transfer. In this case, equation (5-75) is modified to account for enhanced mixing.

For three-dimensional clouds,

$$d(\rho u_e h_e B_e)/dx = 0.4 \, \rho_a w B_e \, (1+\alpha)/\phi(Ri'_b) \qquad (5\text{-}76)$$

where the plume effective half width is defined by

$$B_e = b + 0.5 \, \pi^{1/2} \sigma_y \qquad (5\text{-}77)$$

The coefficient b and the lateral $\sigma_y$ are discussed in Colenbrander (1980). The parameters w and $Ri'_b$ account for heat transfer and are explained in detail by Havens and Spicer (1985). The lateral spread of the cloud is modeled by

$$dB_e/dx = u_e^{-1} \, dB_e/dt = u_f/u_e \qquad (5\text{-}78)$$

The average transport velocity in the plume is defined by

$$u_e = \int_0^{\infty} C u \, dz / \int_0^{\infty} C dz = u_1 (\sigma_z/z_1)^{\alpha} / \Gamma((1+\alpha)^{-1}) \qquad (5\text{-}79)$$

where subscript 1 indicates a reference height.

The crosswind similarity parameter $\sigma_y(x)$ is also determined by requiring that it satisfy the diffusion equation with a specified horizontal eddy diffusivity (see Havens and Spicer 1985 for details). For a steady plume, the centerline concentration $C_C$ is determined from the material balance

$$Q = \int_0^\infty \int_{-\infty}^\infty Cudydz = 2C_C(u_1 z_1/(1+\alpha))(\sigma_z/z_1)^{1+\alpha}B_e \qquad (5\text{-}80)$$

where $Q$ is the plume source strength.

For some simulations of cryogenic gas releases, heat transfer to the plume in the downwind dispersion calculation may be important, particularly in low wind conditions. The source calculation determines a gas/air mixture initial condition for the downwind dispersion problem. Air entrained into the plume is assumed to mix adiabatically. Heat transfer to the plume downwind of the source is also important and is determined by an energy balance on a uniform cross-section.

Following the HEGADAS model, an adjustment to the value of $C_C$ is applied to account for dispersion parallel to the wind direction. As is typical of other components of DEGADIS, this adjustment is made using empirical coefficients found in the literature. There are several unique empirical components of this model, such as the "source blanket" and "moving observer" concepts. For discussion of this and other details the reader is referred to Havens and Spicer (1985).

<u>FEM3</u>

The FEM3 model (Chan 1983) is chosen to be discussed in this section because it is representative of three-dimensional models and is very completely described in the reference. Equations (5-39) through (5-49) are the governing equations used by FEM3. They are solved simultaneously at nodes on a three dimensional grid which is fine enough that several grid points are included in the initial cloud. As mentioned earlier, a major difficulty with this type of model is the specification of the eddy diffusivity (K) coefficients. They are functions of the local wind field and stability, but these variables are altered by the presence of the dense gas cloud. Furthermore, the eddy-diffusivity approach is valid only as long as the cloud is larger than the length-scale of the turbulent eddies.

The numerical procedure used to solve the governing equations must be very cleverly constructed in order to avoid the generation of short-period waves, pseudo-diffusion, and other numerical instabilities. The presence of non-linear terms such as $\underline{u} \cdot \nabla \underline{u}$ in the governing equations is the main reason for these problems, which receive more discussion in reports on numerical models than other aspects of the model (e.g. Chan et al. 1981). The FEM3 model uses the method of spatial discretization of the governing equations by the finite element method in conjunction with the Galerkin method of weighted residuals. The methods are described in great detail by Chan et al (1981).

The code is capable of producing voluminous output, since all dependent variables are predicted in time and at hundreds of spatial grid points. Also, the code requires a relatively long time to execute on a supercomputer. Figure 5-16 contains an example (from Ermak et al 1982) of the cross-wind velocity field produced by the model in simulating two experiments in the Burro series. The strong lateral velocity near the ground at the dense gas front can be seen, as well as return flow aloft and vertical motions throughout the cloud.

### INPUFF-2.0

The INPUFF model represents the class of puff trajectory models used for non-dense emissions (Petersen and Lavdas 1986). They are not expected to be valid for very dense gases close to the source, but are useful for many types of hazardous gas releases with nearly-neutral or slight positive buoyancy. Although the authors do not mention any specific density criteria for application of their model, the critical Richardson number criterion mentioned earlier would be relevant (i.e. if $Ri_o \ll 1$, the gas behaves as a neutrally - buoyant gas and INPUFF could be used). Plume rise from positively buoyant plumes is accounted for by the model using standard formulas found in EPA UNAMAP models.

The model is very straightforward, employing the Gaussian puff equation (5-54 and 5-55) to determine the distribution of concentrations in each puff. Dispersion coefficients derived for plumes are used because they are based on extensive experimental evidence. The rate of dispersion can change with time

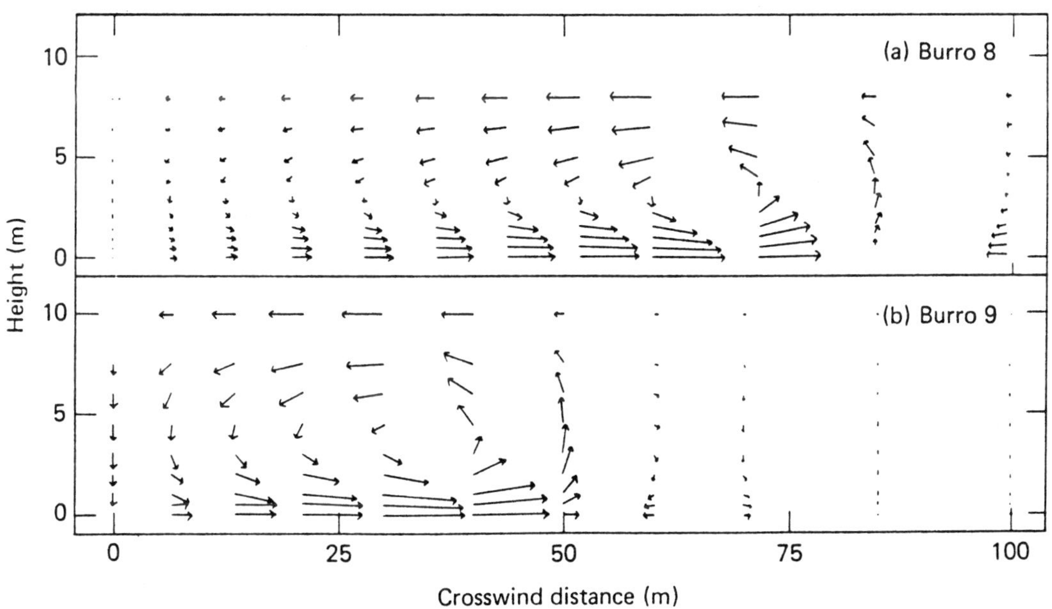


Figure 5-16.   Velocity predictions (arrows) by the FEM3 numerical model,
for two experiments in the Burro series (Ermak and Chan 1985).
The figures represent a cross-section through half of the
plume, with the plume centerline on the left edge.

as atmospheric stability changes. Multiple sources can be handled by INPUFF. It is assumed that puffs are released from each source for each time increment, where the time increments are typically 15 minutes to one hour. The observed wind field is used to advect each puff, and the predicted concentration at any receptor at any time is obtained by summing the contribution from individual puffs. An example of a possible arrangement of meteorological and receptor grids is given in Figure 5-17.

INPUFF can use on-site turbulence ($\sigma_v$ and $\sigma_w$) to calculate the dispersion coefficients:

$$\sigma_y = \sigma_v t f_y = \sigma_x \tag{5-82}$$

$$\sigma_z = \sigma_w t f_z \tag{5-83}$$

The parameters $\sigma_v$ and $\sigma_w$ are the standard deviations of the turbulent fluctuations in lateral and vertical wind speed, respectively. The dimensionless functions $f_y$ and $f_z$ are given by

$$f_y = (1 + 0.9 \, (t/1000\text{sec})^{1/2})^{-1} \tag{5-84}$$
$$f_z = 1 \qquad \text{unstable conditions} \tag{5-85a}$$
$$f_z = (1 + 0.9 \, (t/50 \text{ sec})^{1/2})^{-1} \quad \text{stable conditions} \tag{5-85b}$$

For many applications, the travel time is long enough that the puff disperses vertically through the mixed layer. If the mixing depth then collapses, as it does most evenings, the model assumes that the puff remains well-mixed through the mixed layer, h, that existed prior to collapse.

### 5.2.2 Summary of Models

Appendix A contains the results of questionnaires sent to model developers. That section tends to emphasize models that have been developed for sale to industries and government agencies. In many cases the model involves a package of software and hardware for use in training personnel or planning for evacuations. However, other models are available for specific components of the problem, such as the entrainment rate or the generation of aerosol. The purpose of the present subsection is to tabulate some various

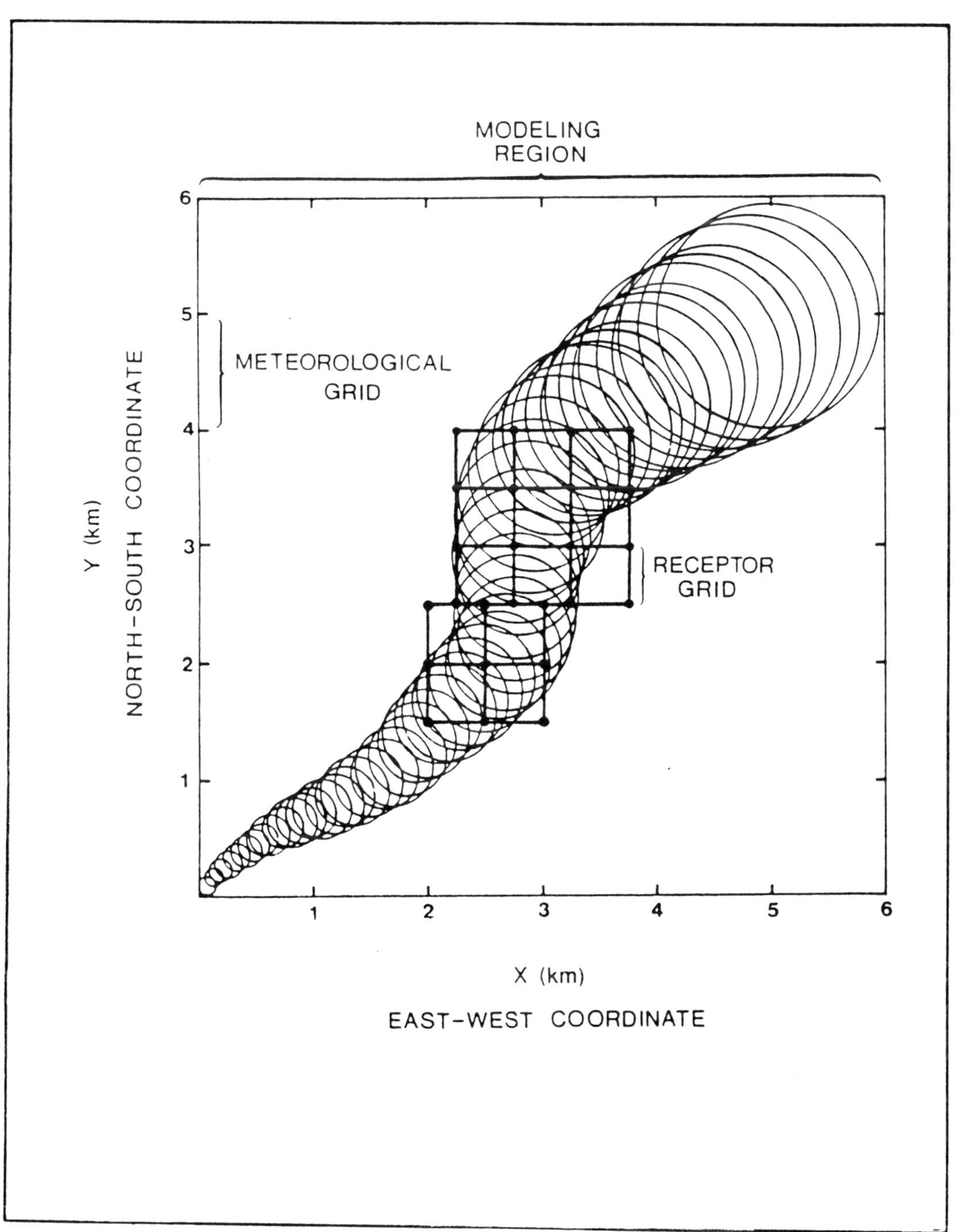



Figure 5-17.  Example of meteorological and receptor grids for application of INPUFF (from Petersen and Lavdas 1986).

111

available modeling systems and components, for quick reference by the reader. This is done in Table 5.4 in alphabetic order, with the capabilities of 40 models checked on a matrix. The list of capabilities is kept short to enhance readability. In most cases the "model" is discussed in more than one reference (e.g HEGADAS is presented in a number of papers by Puttock and Colenbrander). Source emission model capabilities are not considered here since they were covered in Section 4 and will be further discussed in Appendix A.

An instantaneous puff refers to a case where the duration of release is much less than the averaging time or the travel time from the source to the receptor. A continuous plume refers to a case where the duration of release is much greater than the averaging time or the travel time from the source to the receptor.

EPA and NRC regulatory models for neutrally or positively buoyant gases are not included here because they are extensively discussed in other review documents (e.g. Hanna, Briggs and Hosker 1982). Also, if a researcher makes only a slight modification to an existing model, his new model will probably not appear in the table. Some older models do not appear either because they have been replaced by improved models developed within the same organization, or because they are no longer in use.

Table 5.4

List of 40 Available Models for Hazardous Gas Transport and Dispersion.
See Appendix A for More Details on Some of the Models.  An x in a
Column means that the Model Treats that Particular Phenomenon.

| Model | Reference | Instantaneous Puff | Continuous Plume | Dense Gas | Slab Similarity | Numerical (Grid Model) | Winds Vary in Space |
|---|---|:---:|:---:|:---:|:---:|:---:|:---:|
| AIRTOX | Paine et al 1986 | x | x | x | x | | |
| AVACTA II | Zannetti et al 1986 | x | x | | x | | x |
| Britter | Britter 1979, 1980 | | x | x | x | | |
| CARE | Verholek 1986 | x | x | x | x | | x |
| CHARM | Radian 1986<br>Balentine and<br>Eltgroth 1985 | x | x | x | x | x | x |
| Chatwin | Chatwin 1983 | x | | x | x | | x |
| CIGALE 2 | Crabol et al 1986 | x | | x | x | | |
| COBRA III | Oliverio et al 1986 | x | x | x | x | | |
| CRUNCH | Fryer 1980, Jagger 1983 | | x | x | x | | |
| 3D-MERCURE | Riou and Saab 1985 | x | x | x | | | x |
| D2PC | Whitacre et al 1986 | x | x | | x | | |
| DEGADIS | Havens and Spicer 1985 | x | x | x | x | | |
| DENS20 | Meroney 1984 | x | x | x | x | | |
| DENZ | Fryer and Kaiser 1979 | x | | x | x | | |
| Fay & Zemba | Fay and Zemba 1985,1986 | x | x | x | x | | |
| FEM3 | Chan 1983 | x | x | x | | x | x |
| HAZARD | Drivas et al 1983 | x | x | x | x | x | x |
| HEAVY GAS | Deaves 1985 | x | x | x | | x | x |
| HEAVYPUFF | Jensen 1983 | x | | x | x | | |

Table 5.4 (continued)

| Model | Reference | Instantaneous Puff | Continuous Plume | Dense Gas | Slab Similarity | Numerical (Grid Model) | Winds Vary in Space |
|---|---|---|---|---|---|---|---|
| HEGADAS | Colenbrander and Puttock 1983 | x | x | x | x | | |
| HEGDAS | Morrow 1982 | | x | x | x | | |
| Hoot et al | Hoot et al 1973 | | x | x | x | | |
| INPUFF 2.0 | Petersen & Lavdas 1986 | x | | | x | | x |
| MADICT | Ludwig 1985 | x | | | x | x | x |
| MESOPUFF II | Scire and Lurmann 1983 | x | | | x | | x |
| MIDAS | Woodward 1987 | x | x | x | x | | x |
| NILU | Eidsvik 1980 | x | | x | x | | |
| Ooms et al | Ooms et al 1974 | | x | x | x | | |
| Port Comp. System | MOE 1983, 1986 | x | x | x | x | | |
| RIMPUFF | Mikkelsen et al 1984 | x | x | | x | | x |
| SAFEMODS | Raj 1981, 1986 | x | x | x | x | | |
| SAFER | Personal Correspondence | x | x | x | x | x | x |
| SIGMET | England et al 1978 | x | x | x | | x | x |
| SLAB | Morgan et al 1983 | x | x | x | x | | |
| SPILLS | Fleischer 1980 | x | x | | x | | |
| TOXGAS | McCready et al 1986 | x | | | x | | |
| van Ulden | van Ulden 1974 | x | | x | x | | |
| VAPID | Jensen 1983 | x | x | x | x | | |
| Webber | Webber and Brighton 1986 | x | | x | x | | |
| Wilson | Wilson 1981 | x | x | | x | | |
| Zeman | Zeman 1982 | x | x | x | x | | |

# 6

# Model Evaluation Experience and Uncertainties Estimates

It is easy to develop a model, but more difficult to demonstrate that it agrees with observations.  Havens (1978) showed that the few dense gas dispersion models available in 1978 produced concentrations that varied over one to two orders of magnitude.  Even the best of models show considerable uncertainty due to input data errors and the stochastic nature of the atmosphere.  This section describes some field experiments on hazardous gas dispersion, some results of model evaluations with these data sets, and some examples of the uncertainties involved.

6.1  Summary of Data Sets

The greatest interest is in full scale atmospheric field experiments. Unfortunately, data are very sparse for serious accidents, such as those mentioned in the Introduction.  For example, photographs of the Houston tanker crash, in which 19 tons of anhydrous ammonia were released, suggest that the initial cloud height was 35m, and that, after one minute, the cloud was 300m wide and 600m long (Kaiser and Walker 1978).  However, no in-cloud concentration observations were made.  In the same reference it is stated that, in the Potchefstroom ammonia incident, "(the model prediction) is consistent with the positions of those found dead after the accident."  The physical data required for comprehensive model evaluation are not available from any reported accident.

At the other extreme, there have been a number of heavily-instrumented dense gas dispersion experiments performed in laboratory wind tunnels and water channels (see Section 5.1.4).  However, these laboratory experiments generally stress one specific aspect of the problem and ignore others, such as ambient stratification, mesoscale eddies, heat exchanges, and entrained aerosols.  As a result, limited field experiments were begun in the 1960's.  By the 1980's, several well-funded experiments were underway.  Puttock, Blackmore and Colenbrander (1982) review the field experiments through about 1980, and Puttock and Colenbrander (1985) review more recent field experiments.  Most of the dense gas field experiments prior to 1980 involved small quantities of LNG spilled onto water surfaces or into small diked areas.  In the 1980's, the field experiments were expanded in scope to include larger releases of other gases besides LNG.  Table 6-1 contains a summary of some of the more useful field experiments, in order to give the reader an overview of the types of data available.  The reader is referred to Puttock et al (1982) and Puttock and Colenbrander (1985) for a more detailed discussion of a more extensive list of experiments.  The reports and papers on the individual experiments will provide some of the details needed for model evaluation exercises (some specific results are summarized in the next subsection).  In most cases, it is necessary to contact the authors to obtain complete data sets.

The field experiments in the table are limited to dense gas dispersion, and data are generally taken only in the region close to the source where the cloud is dominated by gravity slumping.  When gases such as ammonia or $N_2O_4$ are studied, it is found that the generation of aerosol is an important parameter which is not yet satisfactorily measured or modeled.

6.2  Examples of Model Evaluations

During the past five years, a large number of studies have been published in which the predictions of hazardous gas dispersion models are evaluated using the results of the recent comprehensive field experiments reviewed in the last subsection.  In most cases, these models and experiments deal with the gravity slumping phase of the problem.  Also, in most cases, only rudimentary statistical measures are employed and confidence limits are not calculated.  Of the dozens of reports on this subject that have been published, 12 examples are listed in Table 6.2.  Some of these will be reviewed in more detail below.  It is evident in the table that the Thorney Island (Freon), Burro (LNG), Coyote (LNG), Eagle ($N_2O_4$), Maplin Sands (LNG), and Desert Tortoise ($NH_3$) data sets are the most often used in model evaluation.  The layout of receptors in the Eagle and Maplin Sands experiments are reproduced in Figures 6.1 and 6.2 in order to show the number and distance range of the receptors.

The EPA has published a set of suggested model evaluation procedures (EPA 1984) and has sponsored numerous applications of these procedures (e.g. Londergan et al 1983) to neutral or positively buoyant releases from continuous sources.  Other researchers have further refined these methods and proposed

Table 6.1
Examples of Dense Gas Field Experiments
(Adapted from Puttock et al 1982 and Puttock and Colenbrander 1985, plus more recent material)

| Site and Reference | Material | No. of Tests | $Q_I$ ($m^3$ liquid) | $Q$ ($m^3 min^{-1}$ liq) | Surface |
|---|---|---|---|---|---|
| DGA Netherlands<br>van Ulden 1974 | Freon 12 | 2 | 1000 kg | – | sand |
| Gaz de France<br>Humbert and Montet 1972 | LNG | 40 | – | 0.16 | soil |
| ESSO/API<br>Feldbaner et al 1972 | LNG | 17 | .09-10.2 | – | sea |
| HSE Porton<br>Picknett 1981 | Freon | 35 | $40m^3$ gas | – | grassland |
| Maplin Sands<br>Puttock et al 1982 | LNG, propane | 34 | 27 | 1-5 | sand, sea |
| China Lake Burro<br>Koopman et al 1982 | LNG | 8 | 40 | 12-18 | pond |
| China Lake Coyote<br>Goldwire et al 1983 | LNG | 15 | 3-28 | 6-19 | pond |
| Desert Tortoise<br>Koopman et al 1984 | $NH_3$ | 4 | 60 over 7.5 min. | | sand |
| Eagle<br>McRae et al 1983 | $N_2O_4$ | 6 | 4.2 over 3 min. | | sand |
| Thorney Island<br>Puttock and Colenbrander<br>1985 | Freon | 28 | 2000 instantaneous | | airport |

Table 6-2

Some Examples of Hazardous Gas Dispersion Model Evaluations

| Authors | Models | Data Sets |
|---|---|---|
| Layland et al. 1986 | INPUFF 2.0, DEGADIS, OME, PUFF | Eagle ($N_2O_4$), Thorney Is. (Freon) |
| Paine et al. 1986, Heinold et al. 1986 | AIRTOX | Frenchman Flat ($NH_3$), Thorney Is. (Freon), Burro (LNG), Coyote (LNG) |
| Ermak et al. 1982, Ermak and Chan 1985 | Gaussian, SLAB, FEM3 | Burro (LNG), Eagle ($N_2O_4$) |
| McRae 1985 | OB/DG, Gaussian, CHARM, SPILLS | Eagle ($N_2O_4$) |
| Alp et al. 1985 | COBRA, HEGADAS | Maplin Sands (LNG) |
| Riou and Saab 1985 | Box, MERCURE-GL | Thorney Is. (Freon) |
| Balentine and Eltgroth 1985 | CHARM | Burro (LNG), Eagle ($N_2O_4$) |
| Lewellen et al. 1985 | MESO models, Gaussian, ADPIC, IMPACT, Others | INEL $SF_6$ data |
| Wheatley et al. 1985, 1986 | Picknett, DENZ | Thorney Is. (Freon) |
| Puttock et al. 1984 Puttock and Colenbrander 1985 | HEGADAS | Maplin Sands (LNG), Thorney Is. (Freon) |
| Fay and Ranck 1983 | Their own model | Porton (Freon), van Ulden data, Thorney Is. (Freon) |
| Spicer et al 1986 | DEGADIS | Burro (LNG), Maplin Sands (LNG) Thorney Is. (Freon), Welker (LPG) |

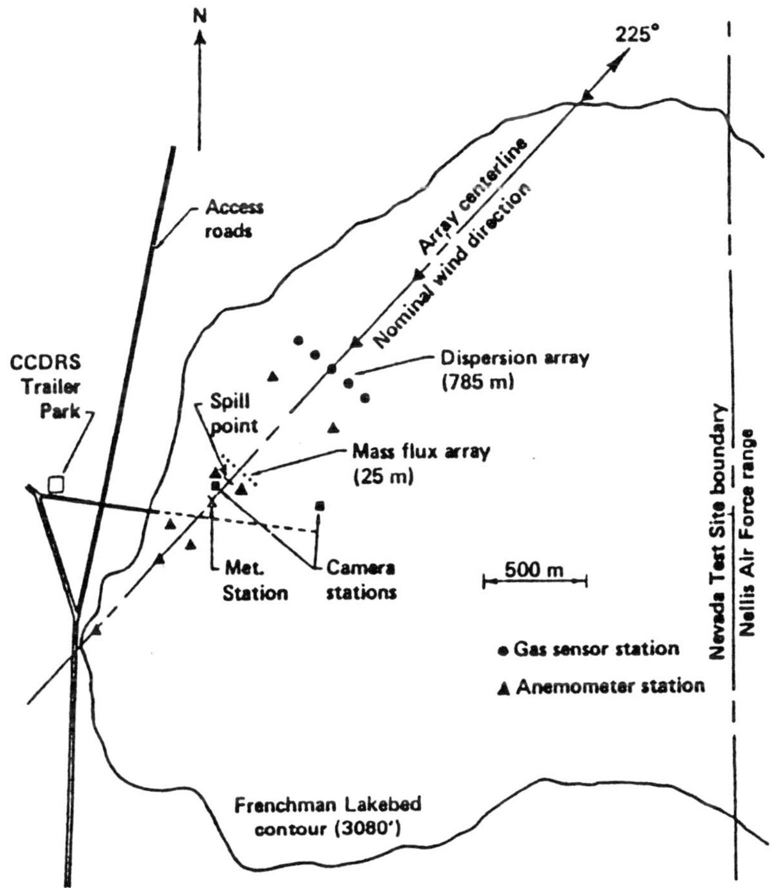


Figure 6-1. Site Diagram for Eagle Series of Dense Gas Dispersion Experiments. (McRae 1985).

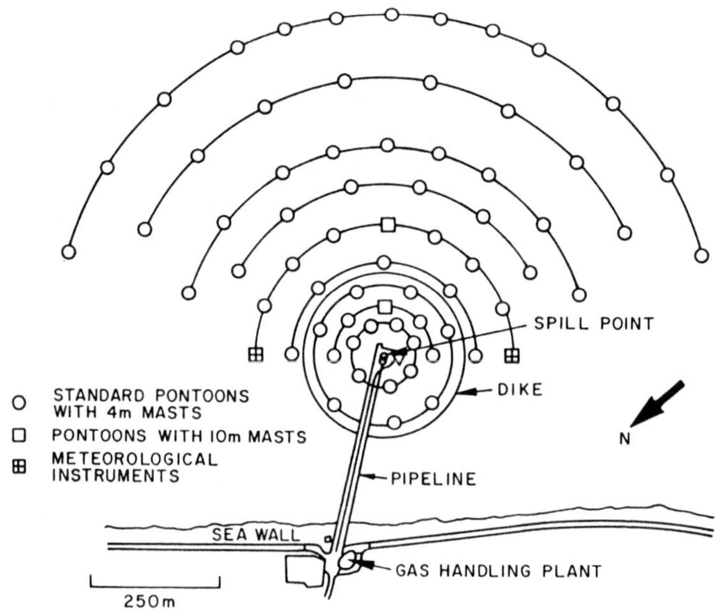



Figure 6-2.   Site diagram for most of continuous LNG spills at Maplin Sands (Puttock et al 1984).

schemes for estimating confidence intervals (e.g. Cox and Tikvart 1985, Hanna 1986). Some typical performance measures that are used by Hanna (1986) are the following

$$\text{bias} = \overline{C}_p - \overline{C}_o \qquad\qquad (6\text{-}1)$$

$$\text{mean square error} = \overline{(C_p - C_o)^2} \qquad\qquad (6\text{-}2)$$

$$\text{correlation } r = \overline{(C_p - \overline{C}_p)(C_o - \overline{C}_o)}/\sigma_{Cp}\sigma_{Co} \qquad\qquad (6\text{-}3)$$

percent within a factor of two

where $C_p$ is the prediction, $C_o$ is the observation, and an overbar indicates an average. The non-parametric bootstrap resampling procedure (Efron 1982) is often used to estimate confidence intervals. In light of these developments, it is surprising that none of the model evaluation studies in Table 6-2 use the EPA procedures. Instead, a typical study contains plots of observed and predicted ground level concentration contours, graphs of peak concentration versus distance, or tables of observed and predicted distance to the lower flammability level (LFL). This is an area where application of the advanced procedures would greatly enhance the results of the comparisons.

A few specific results of the studies are reviewed below.

Layland et al (1986)

Concentration versus distance is plotted for two Thorney Island trials in Figure 6-3. The DEGADIS model predictions are closest to the observations, although the model shows a discontinuity at a distance of about 100 to 300m, as the model shifts internally from one algorithm to another. The PUFF and INPUFF models do not account for puff excess density and are seen to be sometimes wildly wrong (plus or minus an order of magnitude).

Heinold et al (1986)

The AIRTOX model comparisons are as comprehensive as any in Table 5.2, since four data sets are used. Unfortunately, results are given for only one model. It is found that the range of the mean ratio $C_p/C_o$ for the several experiments is 0.7 to 3.4, with a median value of 1.0. The ratio $C_p/C_o$ is plotted in the reference as a function of downwind distances, showing a

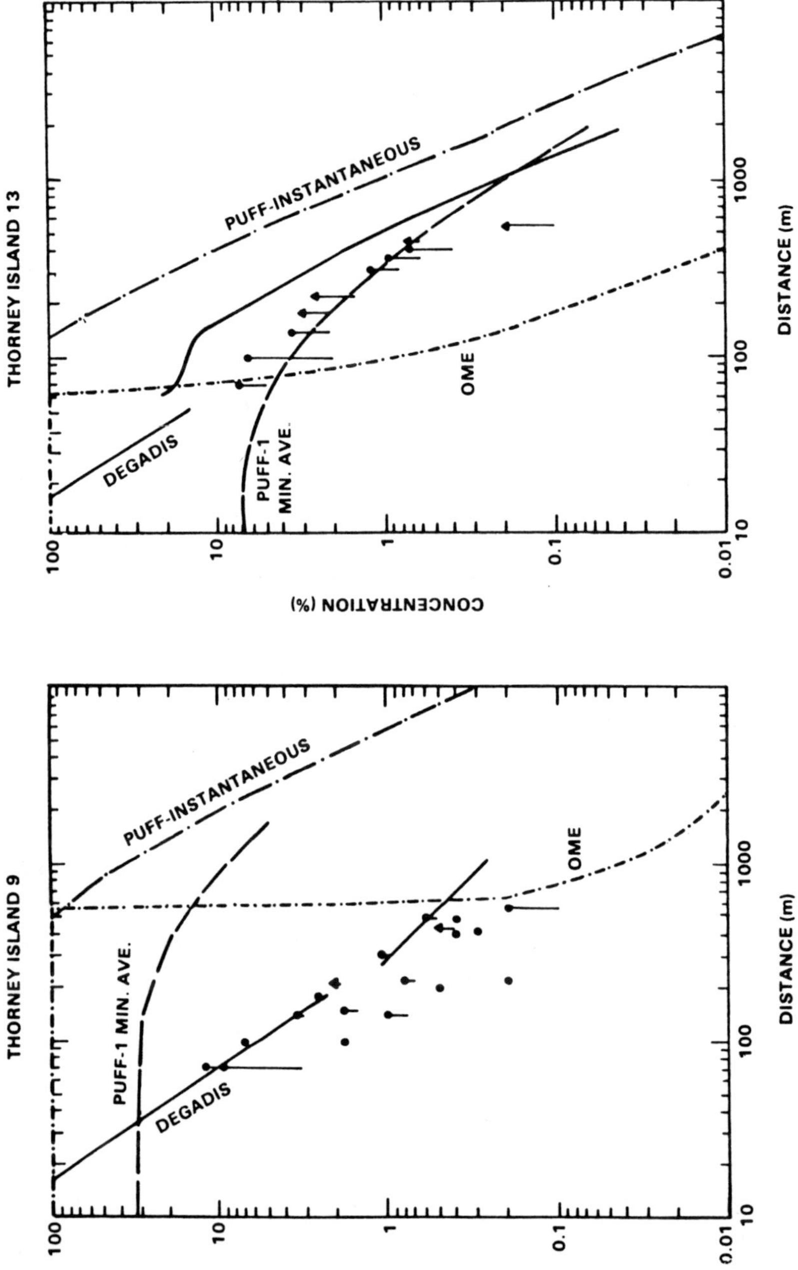


Figure 6-3. Observations and model predictions of concentrations as a function of distance (from Layland et al 1986).

122

tendency to overpredict at small distances ($x < 200$m).  The standard deviation of $C_p/C_o$ at any distance can be estimated to be about 30%.

<u>Ermak et al (1982, 1985)</u>

The SLAB and FEM3 model predictions are compared with Burro and Eagle observations using plots of observed and predicted ground level contours and graphs of concentration versus distance.  The reader generally is required to interpret these plots on a qualitative basis.

<u>McRae (1985)</u>

The peak $NO_2$ concentration at a distance of 785m in the Eagle experiments is used as a basis for comparison of four models.  All models underpredict by a factor of two or three, on the average.

<u>Alp et al (1985)</u>

The COBRA and HEGADAS model predictions of the distance to the 5% LFL at the Maplin Sands site are compared with observations using the statistics:

$$MRE \text{ (mean relative error)} = \overline{(x_p - x_o)/x_o} \tag{6-4}$$

$$MARE \text{ (mean absolute relative error)} = \overline{|x_p - x_o|/x_o} \tag{6-5}$$

It is found that MRE equals $-.12$ and $+.22$ for COBRA and HEGADAS, and MARE equals .24 and .38 for the two models.

<u>Balentine and Eltgroth (1985)</u>

The CHARM model is shown to be capable of simulating the observed Burro and Eagle experiment plume widths and heights within about $\pm 50$%.  Figure 6-4 contains a comparison of the observed and predicted plume cross-sections for Burro 9, showing that the basic structure of the plume is well-simulated.

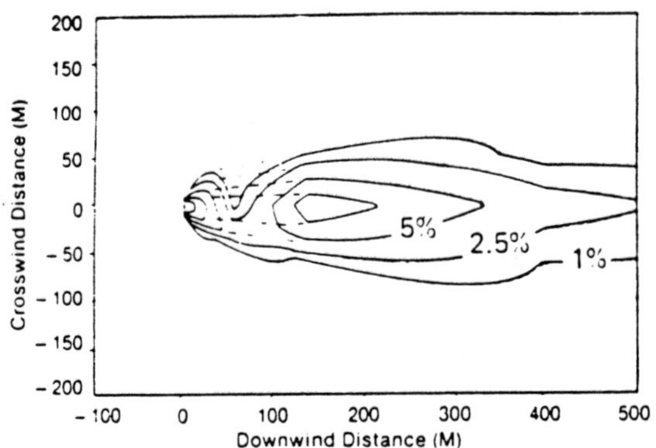


Burro 9 observed horizontal plume cross-section
for 80 seconds downwind.  Concentrations are methane percent
by volume.  The dotted lines represent an attempt by LLNL to
correct for the effect of explosions on the methane detectors.

Source (Koopman, et al, 1981)

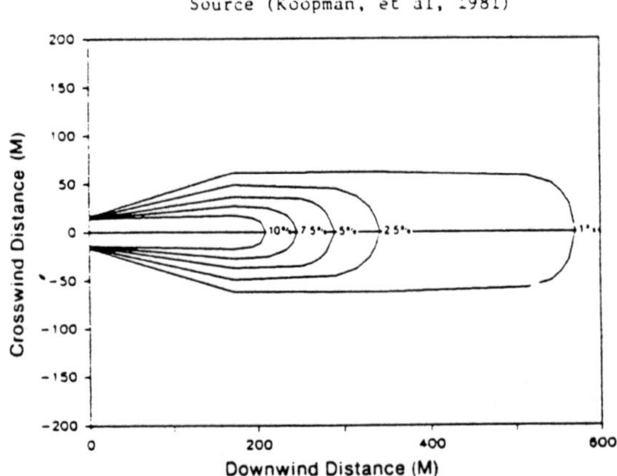


Burro 9 horizontal plume cross-section predicted by CHARM for 90 seconds downwind.  Concentrations
are methane percent by volume.

Figure 6-4.  Observed and predicted Burro 9 horizontal plume cross-section
(from Balentine and Eltgroth 1985).

<u>Lewellen et al (1985)</u>

The models and data in this report all refer to a neutrally-buoyant gas. The results are quite discouraging for developers of complex models, since analyses of the observed and predicted plume contours at the Idaho National Engineering Laboratory suggest that the simple Gaussian straight-line models work just as well as three-dimensional numerical models.

<u>Wheatley et al (1985)</u>

Some characteristics of the Thorney Island data (e.g. cloud height and concentration) are compared with predictions of the DENZ model and a generalized form of the Picknett model. A Goodness of Fit Measure (GFM) is used which is a weighted sum of squares of deviations of data points from a possible fit. Model parameters (e.g., entrainment constant) are chosen such that GFM is minimized.

<u>Puttock et al (1984, 1985)</u>

Again, the primary interest of this study is the distance to the LFL, which can be predicted consistently within a factor of two by the HEGADAS model, when compared with the Maplin Sands data. An example of observed and predicted concentrations as a function of distance is shown for Spill 15 in Figure 6-5.

<u>Spicer et al (1986)</u>

The DEGADIS model is compared with observations at four experiments. Because the model as described in the reference is intended for application to LNG, a comparison of distances to 1, 2.5, and 5% LFL are given. It is found that 90% confidence limits on the ratios of observed to predicted distance for the three LNG experiments are .95 to 1.24 for a 1% LFL, .82 to 1.03 for a 2.5% LFL, and .73 to .96 for a 5% LFL. The ratio $C_o/C_p$ for the Eagle $NO_2$ experiment is about unity for the DEGADIS model and about ten for the Ocean Breeze-Dry Gulch (OBDG) model. Of course the OBDG model is valid only for neutrally-buoyant gases.

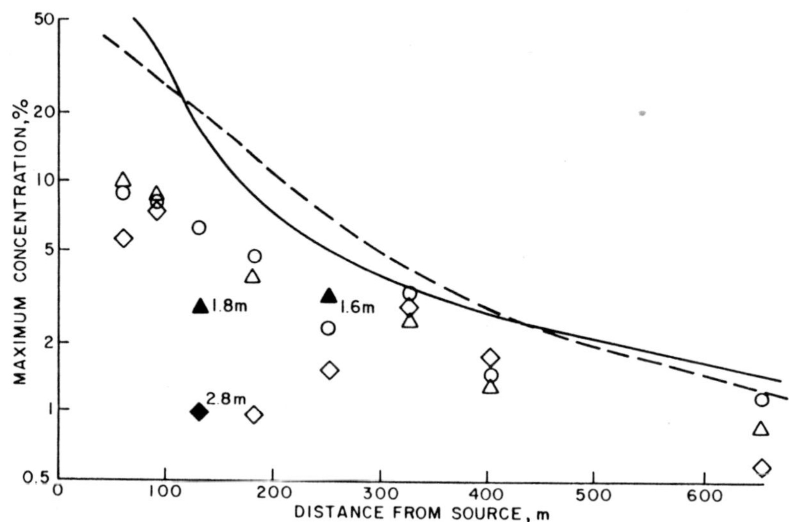


Figure 6-5.  Maximum concentrations (smoothed using 3 s average) measured
at each radius for spill 15, at heights:  O 0.9-1.0m,
1.3-1.5m,  2.2-2.4m.  HEGADAS II predictions at the surface:
___________ "pool" evaporation and -----"jet" evaporation.
(From Puttock et al 1984)

These model evaluation studies represent a vast improvement over the situation noted by Havens in 1978. But it is clear that there is much yet to do both in extending the range of the experiments and in the applications of a more complete set of evaluation statistics.

## 6.3 Uncertainties

If major decisions are to be made regarding pollution control equipment, evacuation plans, and risk assessments regarding hazardous gases, it is important to have the best possible information on our confidence in the models that are used and the data that are being collected. It may even be possible to build the confidence intervals (uncertainty) into the decision-making process. The purpose of this subsection is to discuss the three components of total error or uncertainty in models for source emissions, transport, and dispersion of hazardous gases.

Errors caused by model physics assumptions

Random variability (turbulence)

Errors generated by input data errors

These components have not yet been studied in any comprehensive way. Our general philosophy on model evaluation and uncertainties is shown in Figure 6-6, taken from Hanna (1986), where the three components of uncertainty are plotted as a function of the number of parameters in the model. It is desirable to construct a model such that the total model uncertainty on the figure is minimized. The uncertainties can be quantified by defining total uncertainty as $(C_p - C_o)^2$ and assuming that $C_p$ and $C_o$ are given by:

$$C_o = \overline{C_{oa}} + C'_o + \Delta C_o \qquad (6-6)$$

$$C_p = \overline{C_p} + C'_p + \Delta C_p \qquad (6-7)$$

where $\overline{C_{oa}}$ is the actual ensemble average

$C'_o$ is the stochastic (random) variability

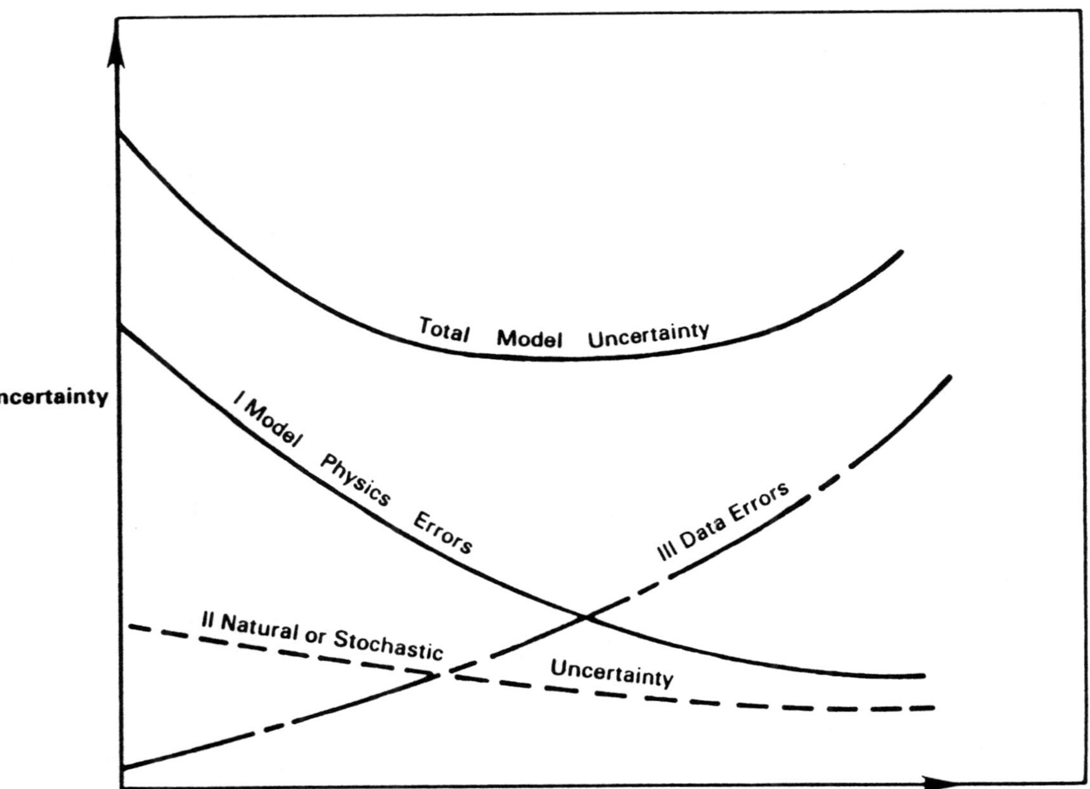


Figure 6-6.   Illustration of variation of model uncertainty components with number of parameters in model.

$\Delta C_o$ is the data error in $C_o$

$\overline{C}_p$ is the predicted ensemble average

$\Delta C_p$ is the error due to data input errors

$C'_p$ is the predicted random variability

No current hazardous gas models predict $C_p'$. If it is assumed that there is no correlation among the components, then the total uncertainty is given by the sum of the three components shown in Figure 6-5.

$$\overline{(C_p - C_o)^2} = \overline{(C_p - C_{oa})^2} + \overline{\sigma^2_{Co}} + \overline{\Delta C'^2_o + \Delta C'^2_p} \tag{6-8}$$

| Total Model Uncertainty | Model Physics Error | Stochastic Uncertainty | Data Errors |
|---|---|---|---|
| | I | II | III |

As shown in Section 5.1.6, the stochastic uncertainty component II is roughly equal to the square of the mean (i.e., $\sigma^2_{Co}/\overline{C}_o^2 \sim 1$) and can be predicted by some models (e.g. Hanna 1984). The data errors component III is also believed to be of the same order as the square of the mean. The model physics error Component I can be estimated by solving equation (6-8) for that component, given the total model uncertainty and components II and III. Of course this procedure is highly uncertain if the components II and III are approximately equal to the total model uncertainty.

The data errors component III can be estimated based on studies of instrument errors in the field. It is stressed that some QC(quality control) procedures, such as checking the voltage output of an anemometer at a given rotation rate, do not tell much about actual instrument error <u>in the field</u>. These actual errors are best determined through use of co-located instruments and comparison with high quality "base-line" instruments. In most field experiments, this option is not practical. Information on data errors was obtained using colocated instruments as part of the tracer study of dispersion from tall stack plumes, sponsored by the Electric Power Research Institute (EPRI). Table 6-3 gives some figures on uncertainties for some instruments

Table 6-3

Uncertainties in EPRI Aerometric and Meteorological Data, Kincaid Site
(from Hanna et al 1986)

| Parameter | Standard Deviation or Other Limit |
|---|---|
| Source $SF_6$ | 3% |
| Monitored $SF_6$ | 6% for C > 100 ppt, 6 ppt for C < 100 ppt |
| Mixing height h | 20% (day), 100% (night) |
| Wind direction to max. observed C | $20^{\circ}$ |
| Friction velocity $u_*$ | 40% |
| Wind speed u | 10% |
| Vertical temperature gradient d /dz | $0.01^{\circ}C/m$ |

(from Hanna et al., 1986).  The wind direction error is obtained by calculating the difference between the observed wind direction and the direction from the source to the receptor with the maximum observed concentration.  The site consists of flat farmland with some man-made lakes interspersed in the area, and the instruments were all "off the shelf" and operated by technicians rather than research engineers.  Thus these data errors are typical of what would be found in a routine monitoring network in the neighborhood of any industrial site.

In the case where these models are used for planning purposes, the data errors could also be interpreted as being due to poor approximations of test scenarios.

The $\overline{\Delta C'_o}^2$ term in equation (6-8) is simply the error in the concentration measurement.  If there is a minimum error of 6% in the observation of C, as suggested in Table 6-3, then the total model error in a model evaluation exercise clearly has a minimum of 6%.  If the observed concentrations in the EPRI study had an average of 12 ppt, this total model uncertainty would grow to 50%, since the data error is 6 ppt at low concentrations.  For example, the model can always be written in the form:

$$C = f(x_1, x_2, x_3, \ldots x_n)$$

Then the term $\overline{\Delta C'_p}^2$ in equation (6-8) can be calculated from:

$$\overline{(dC/C)}^2 = ((\partial f/\partial x_1)(dx_1/x_1))^2 + ((\partial f/\partial x_2)(dx_2/x_2))^2 + \ldots \qquad (6-9)$$

where it is assumed that the dx terms are not correlated.

As an example of the use of this method, assume a simple formula for pollutant concentration:

$$C = AQ/u \qquad (6-10)$$

where A is a constant.  Then equation 6-9 yields:

$$\overline{(dC/C)}^2 = \overline{(dQ/Q)}^2 + \overline{(du/u)}^2 \qquad (6-11)$$

If the observed data errors from Table 6-3 are used for dQ and du, then the expected error in the predicted concentration in $((3\%)^2 + (10\%)^2)^{1/2}$, or 10.4%.  The reader can quickly see that the total error for a model with many data inputs can grow to be quite large.  Some of the dense gas models that include evaporative emissions terms can easily have errors in predictions of 50 to 100% or more simply due to uncertainties in the input.  This aspect of hazardous gas modeling deserves further study in order to estimate our confidence in the predictions.

# 7

# Research Needs

There are dozens of models in existence for modeling accidental releases of hazardous materials to the atmosphere. Nevertheless, this field is still in its infancy and there are several areas that require much more work, such as the modeling of two phase flows and the quantification of aerosol formation. The purpose of this document is to briefly summarize the state-of-the-art, emphasizing simple explanations of basic physical principles and review of available models. The following summary of research needs may be useful for planning future work.

1)    Source Emissions Models. Models exist for calculating emissions from the following source types:  single or binary component liquid spills, liquid jets, and gas jets. However, these models have been subjected to only limited testing and uncertainties are not well known. Preliminary models are under development for calculating emissions from the following source types: multicomponent liquid spills and two phase jets. Recent field studies show that many emissions are in the form of two phase (liquid plus gas) jets but only a few very specialized research models can handle them. New research programs are needed to develop theoretical models and data for all types of potential emissions.

2)    Transport and Dispersion Models. Many models exist for calculating the dispersion of dense gases from elevated point sources or ground level area or volume sources. The best of these models can predict concentrations near the source for very simple source scenarios to within a factor of two. However these models do not handle all the complex thermodynamic processes (e.g., phase changes of liquid pollutant or water) that can occur in a hazardous gas plume containing aerosols. The subject of heat and moisture exchange with the underlying surface requires more study. None of the models has been subjected to a rigorous statistical evaluation with field data. Future research should emphasize the development of comprehensive model approaches that account for all these thermodynamic complexities and the acquisition of field data to aid in model development and testing.

3)    <u>Concentration Fluctuations</u>.  Only a few models account for concentration fluctuations, and then only in a rudimentary way.  Questions of flammability and short term toxicity require the development of improved methods of estimating extreme concentrations based on model predictions of the mean and the variance.  The effects of finite sampling volumes and sampling times must be included.

4)    <u>Input Data Uncertainties</u>.  Modelers and field personnel should work together to develop adequate data sets for model input.  For example, the heat conductivity of the ground must be known to calculate dispersion of a cold cloud.  Uncertainties in model input parameters should be quantified and the effects of these uncertainties on the model predictions assessed.

5)    <u>Field Experiments</u>.  Recent field experiments on simplified dense gas dispersion have been very useful.  More of these experiments in a variety of settings (including complex terrain and sites where obstructions are present) should be conducted.  The greatest need is for field experiments of tank and pipeline ruptures in which the initial acceleration is important or in which two phase (gas plus liquid) flows are generated.  Experiments should also emphasize non-dense gases.  Evaporation from multicomponent spills also requires field study.  The experiments should be carried out to distances where concentrations drop to levels of toxic concern.

6)    <u>Model Evaluation</u>.  Comprehensive means of evaluating air quality model performance (including confidence limits) are now available and should be applied to hazardous gas models.  Up to this point, hazardous gas model evaluations have been very limited.  In any case, a model should be tested with <u>independent</u> observations and its uncertainty assessed in a variety of scenarios.

# *Appendix A*

# Modeler's Responses to Questionnaires

A.1  Procedure

As a result of the literature review performed during the period when this
book was being written, a list of so-called "operational" or "applied" models
was made.  Additional names were obtained from CCPS committee members, who have
been approached by numerous companies interested in providing models.  Rather
than using the existing literature to describe the models, it was decided to
ask the modelers to describe their own models by completing a four-page
questionnaire.  The questions were decided upon by the CCPS committee during
meetings with the authors of this book, and the questionnaire was sent to 42
modelers.  33 completed questionnaires were received from 25 of these
modelers.  Other modelers replied that they were "no longer in the business",
or that their model was used mainly for research purposes.  A few modelers did
not reply.  Apologies are given to those modelers who inadvertently slipped
through this system.

The following subsections contain an example of the questionnaire and a
large summary table of the results.

A.2  Example of Questionnaire

The questionnaire that is reproduced in this section was selected as an
example mainly because the responses are typed rather than handwritten, and
does not reflect any preference of the authors or the CCPS committee.

A.3  Results of Questionnaire

To save space, the information from the questionnaires has been
transferred to two master tables.  Table A-1 contains the 33 model names, the
addresses of the authors, and selected references for the scientific
description and evaluation of each model.  Table A-2 contains detailed
information from the questionnaires.  In many cases, the authors of the model
also provided the output from test cases and several pages of additional
technical discussion of their model.  The reader should contact the authors of
the models directly to obtain more details.

The information in the tables is provided only as a guide, and it is
pointed out that errors may occur due to misunderstandings on the part of
either the author of the model or the authors of this book.  Note that if the
model does not have a source emission submodel or a transport and dispersion
submodel, then no further questions are answered under that subsection.  Other
blanks in the table indicate that the modeler did not answer that question.
Furthermore, some authors pencilled in the phrase "currently under development"
after some questions.  In future years it is expected that much of this
information must be revised.

1.  Please list your address and the name and telephone number of a contact person for this model.  M. Gary Verholek

2.  For this questionnaire, what is the name of the product being described?

    CARE - Computerized Airborne Release Evaluation System

3.  In what form is your product available?
    Software?  Yes _X_  No ____
    Hardware?  Yes _X_  No ____
    Both?      Yes _X_  No ____

4.  Must hardware and software be purchased as a package?

    Yes ____   No _X_

5.  Would you provide results for a series of test cases to allow comparison of your products with others on the market?

    Yes _X_   No ____  If yes, please attach a description of your test case.

6.  Do you consider your product to be useful mainly for research (   ) or for routine applications ( X )?

7.  Does your product accept real time weather data?

    Yes _X_   No ____

8.  Does your product operate in an interactive mode?

    Yes _X_   No ____

9.  In what manner is data input to the model?
    Hand entry                    Yes _X_ No ____
    Data file memory              Yes _X_ No ____
    A disk or tape input system   Yes _X_ No ____

10. For which of the following objectives is your product most often used?

    (a.)  Planning studies to develop accurate consequence assessment estimates.
    b.    Develop the incentive and justifications for investments intended to reduce risk.
    (c.)  Screen potential risks to decide which require further analysis and risk reduction.

(d.) Train operating personnel in assessing potential consequences of accident scenarios.

(e.) Assist in emergency response decisions, such as whether to evacuate or otherwise warn inhabitants in a given area (real-time applications).

f.   Research.

11. Do you specifically model source emission rate?   Yes  X  No ___

   If yes, does your model account for any of the following characteristics?

   Evaporation of spill liquids                                Yes  X  No ___
   Flashing                                                    Yes  X  No ___
   Multicomponents                                             Yes ___ No  X
   Entrainment of spilled liquid as aerosols                   Yes  X  No ___
   Heat transfer from substrate to vapor cloud                 Yes  X  No ___
   List substrates that are treated: _ground water_____________
   Mass transfer in liquid phase                               Yes  X  No ___
   Evaporation of aerosols                                     Yes ___ No  X
   Gas volume flux from rupture in pressurized containers  Yes ___ No  X
   Condensation of moisture in the vapor cloud                 Yes ___ No  X
   How many chemicals are included in the source emissions code?  _50+_
   Impact of wind on evaporation                               Yes  X  No ___

12. Does your model treat releases as:
   a.   Instantaneous   ____
   b.   Continuous (steady-state) X
   c.   Time-varying  X

13. Does your model treat discharges with:
   a.   Low momentum  X
   b.   High momentum (jet)  ____

14. Have you included models which deal specifically with a:
   a.   dense cloud?   Yes  X  No ___
   b.   neutrally-buoyant cloud?   Yes  X  No ___
   c.   buoyant cloud? Yes ___ No  X

15. Is surface roughness accounted for?   Yes  X  No ___

16. How do you handle terrain?
   a.   Ignored; flat terrain assumed  ____
   (b.) Adjustments made to flat terrain results using empirical relations
   c.   Other, specify  _Coupled to wind field model______________

17. Does your advection model handle space and time variations in wind speed and direction?

   Yes  X  No ____

18. Does your dispersion model treat indoor concentrations resulting from an outside plume?

    Yes _____ No __X__

19. Is dispersion in building wakes included?

    Yes __X__ No _____

20. How would you classify your advection/dispersion model?

    Gaussian ___X___
    Box or Slab _______
    K/numerical _______

21. Is along-wind dispersion treated?     Yes __X__     No _____

22. Is vertical wind shear treated?     Yes __X__     No _____

23. Are chemical reactions treated in the dispersing plume?

    Yes __X__ No _____

24. Is dry or wet deposition modeled?   Yes __X__     No _____

25. What averaging time is assumed in your model?   __15 min.__

26. Are concentration fluctuations included?

    No _X_ . By an explicit model ___ . By an empirical Peak to Mean Ratio ___ .

27. Your dispersion algorithm includes specific physical and chemical information for how many chemicals? ____50+____________

28. Scientific Validity

    On a separate page, please provide specific references for the model components discussed above (e.g., the SPILLS model may be used in your model to calculate evaporation).

29. Evaluation

    a. Has your entire system been evaluated with field data?
       Yes _____ No _X_ . Give information and references on separate page.

    b. List (with references) the components of your model that have been evaluated.

**30.** How are output data presented?   Graphs & Tables

**31.** What are the limits (e.g. distances) of your model?   $>200\mu$

**32.** Operating information.

a)   Number of systems sold   approximately 15

b)   Hardware limitations: >640k RAM, Hard Disk; graphics CRT
IBM PC/XT; HP-9816; VAX, etc.

c)   Link to emergency system? (i.e. alarms, real-time gas monitors,
process data input)   RT link to all thru RS-232 Serial Link.


d)   Do you have an ongoing support system to users?
Yes __X__   No ______

e)   What special features are available that are not covered above?

Table A-1

Models for which Questionnaire was Completed.
Addresses and References are Listed.

| Model | Address | Scientific Validity | Evaluation |
|---|---|---|---|
| AVACTA II | P. Zannetti, Aerovironment, 825 Myrtle Ave., Monrovia, CA 91016 | Zannetti et al. 1986 | |
| CARE | G. Verholek, ESC, 200 Tech. Center Dr., Knoxville, TN 37912 | Verholek 1986 | |
| CHARM | H. Balentine, Radian, P.O. Box 9948, Austin, TX 78766 | Eltgorth and Smith 1983 | Balentine and Eltgorth 1985, McNaughton et al. 1986 |
| COBRA III | E. Alp, CSC, 2 Tippett Rd., Downsview, Ont., Canada M3H2V2 | Alp 1985, Oliverio et al. 1986 | |
| CRUNCH | A. Byrne, SRD, Wigshaw L., Culceth, England WA34NE | Jagger 1983 | |
| DEGADIS | J. Havens, Un. Arkansas, Dept. Ch. Eng., Fayetteville, AR 72701 | Havens and Spicer 1985 | Spicer and Havens 1986 |
| DENZ | Same as CRUNCH | Fryer and Kaiser 1979 | |
| D2DC | Commander, USA, CRDC, Aberdeen Proving Ground, MD 21010-5423 | Whitacre et al. 1986 | |
| EAHAP | J. Cornwell, EAI, P.O. Box 1508, Norman, OK 73069 | | |
| Eidsvik | Y. Gotoas, NIAR, P.B. 64, N-Z001 Lillestrom, Norway | Eidsvik 1981 | Gotaas 1985 |
| Emissions | J. Schroy, Monsanto, 800 N. Lindbergh Blvd., St. Louis, MO 63167 | Wu and Schroy 1979 | |
| EPIDIS | J. Woodward, FSE, 200 Woodport Road, Sparta, NJ 07971 | | |
| FEM3 | S. Chan, LLNL, P.O. Box 808, Livermore, CA 94550 | Chan 1983 | Ermak et al. 1982, 1985 |
| GASP | Same as CRUNCH | Webber 1986, Brighton 1985 | |
| HASTE | A. Puri, ERT, 696 Virginia Road, Concord, MA 01742 | Paine et al. 1986 | Heinold et al 1986, McNaughton et al 1986 |
| HEAVYPUFF | N. Jensen, Riso N.L., DK-4000 Roskilde, Denmark | | |
| HEGADAS | M. Pikaar, Shell, Postbus 162, 2501 AN the Hague, Holland | Colenbrander 1980; Colenbrander and Puttock 1983 | Puttock et al 1982 |
| INPUFF2.0 | W. Petersen, EPA, RTP, NC | Petersen and Lavdas 1986 | |
| MIDAS | K. Woodard, Pickard Lowe & Garrick, 1615 M Street, Wash., DC 20036 | | |
| ModSys | Same as COBRA III | | |
| PLUMEPATH | Same as HEGADAS | Ooms 1972, Ooms et al. 1974 | |
| RIMPUFF | T. Mikkelsen, Riso, N.L., DK-4000 Roskilde, Denmark | | |
| SAFE | P. Ross, AOSTRA, 10010-106 St., Edmonton, Alberta T5J 3L8 | | |
| SAFEMODS | P. Raj, TMS, 99 S. Bedford Street, Burlington, MA 01803-5128 | Raj 1981, 1985, 1986 | |
| SAFER | G. Gelinas, Safer, 5700 Corsa Avenue, Westlake Village, CA 91362 | | |
| SAFETI | R. Cox, Technica, 7/12 Tavistock Sq., London WC1H9LT | Ale and Whitehouse 1986 | |
| SLAB | D. Ermak, LLNL, P.O. Box 808, Livermore, CA 94550 | Ermak et al. 1982 | Ermak and Chan 1985 |
| SPILLS | J. Moser, Shell, P.O. Box 1380, Houston, TX 77001 | Fleischer 1980 | Kunkel 1983 |
| SRI PUFF | F. Ludwig, SRI International, Menlo Park, CA 94025 | | |
| TELJET | Same as SAFETI | | |
| TRAUMA | Same as CRUNCH | Wheatley 1986 | |
| VAPID | Same as HEAVYPUFF | Jensen 1983 | |
| WHAZAN | Same as SAFETI | Kayes 1985 | |

Table A-2
Results from Model Questionnaires (Most Answers are Given as Y=Yes or N=No) as of December 1986

| Question | AVACTA II | CARE | CHARM | COBRA III | CRUNCH | DEGADIS | DENZ | D2DC | EAHAP | Eidsvik | Emissions | EPIDIS | FEM3 | GASP | HASTE |
|---|---|---|---|---|---|---|---|---|---|---|---|---|---|---|---|
| Operating Information | | | | | | | | | | | | | | | |
| Form of model: H=Hardware; S=Software | S | H;S | H;S | S | S | S | S | S | S | H;S | S | S | S | S | H;S |
| Main use: R=Research; A=Applied | R;A | A | A | A | R;A | R;A | R;A | A | R | R;A | A | A | R | R;A | A |
| Operate in interactive mode? | N | Y | Y | N | Y | Y | N | Y | Y | Y | N | Y | N | N | Y |
| Support system? | | Y | Y | | N | Y | N | Y | Y | N | N | | | N | Y |
| Number sold or given away? | 2 | 15 | 80 | | >10 | >10 | >10 | 40 | 5 | | 0 | | | | 6 |
| Link to emergency system? | | Y | Y | | N | N | N | N | N | N | N | | | N | Y |
| Input Data | | | | | | | | | | | | | | | |
| Accept real time weather data? | N | Y | Y | N | N | N | N | N | N | N | N | N | N | N | Y |
| Method of data entry: H=Hand; F=data file memory; D=disk or tape | F | H;F;D | H;F;D | H;F | F | H;F | F | H | H | H | H;F | H;F;D | F | H;F | H;F |
| Source Emissions Model? | N | Y | Y | Y | N | N | N | Y | Y | N | Y | Y | Y | Y | Y |
| Evaporation of Spilled Liquids? | | Y | Y | Y | | | | Y | Y | | Y | Y | Y | Y | Y |
| Flashing | | Y | Y | Y | | | | N | Y | | Y | Y | N | N | Y |
| Multicomponents | | N | N | N | | | | N | Y | | Y | N | N | N | N |
| Entrainment as aerosols? | | Y | Y | Y | | | | N | N | | N | Y | N | N | Y |
| Heat transfer, substrate to cloud? | | Y | N | Y | | | | N | Y | | Y | Y | Y | Y | Y |
| Number of substrates (W=Water; S=Soil) | | W;S | | W;S | | | | | W;S | | W;S | W;S | W;S | W;S | W;S |
| Mass transfer in liquid phase? | | Y | N | N | | | | N | Y | | Y | Y | N | Y | Y |
| Evaporation of aerosols | | N | Y | Y | | | | N | N | | N | Y | N | N | Y |
| Gas flux from container rupture? | | N | Y | Y | | | | N | Y | | N | Y | N | N | Y |
| Condens. of moisture in vapor cloud? | | N | Y | Y | | | | N | Y | | N | Y | Y | N | N |
| Wind influence on evaporation? | | Y | Y | N | | | | Y | Y | | Y | Y | N | N | Y |
| Number of chemicals hardwired (I=input by user) | | 50,I | 83 | 5,I | | | | 17 | 21 | | I | 18,I | I | I | I |
| Transport and Dispersion Model? | Y | Y | Y | Y | Y | Y | Y | Y | Y | Y | N | Y | Y | N | Y |
| Releases treated: I=Instantaneous; C=Continuous; V=Variable | I;C;V | C;V | I;C;V | V | C | I;C;V | I | I;C;V | I;C;V | I;C | | V | I;C | | I;C;V |
| Dense cloud? | N | Y | Y | Y | Y | Y | Y | N | Y | Y | | Y | Y | | Y |
| Jet? | Y | N | Y | N | N | N | N | Y | Y | N | | N | N | | N |
| Neutral cloud? | Y | Y | Y | N | Y | Y | Y | Y | Y | N | | Y | Y | | Y |
| Buoyant cloud? | Y | N | Y | N | N | N | N | Y | Y | N | | N | N | | Y |
| Surface roughness? | Y | Y | N | Y | Y | Y | Y | Y | Y | Y | | Y | Y | | Y |
| Complex terrain handled? | Y | Y | N | N | N | N | N | N | N | N | | N | Y | | N |
| U variations in time and space? | Y | Y | N | N | N | N | N | N | N | N | | N | Y | | N |
| Indoor concentrations? | N | N | N | N | N | N | N | N | N | N | | N | N | | N |
| Building wake effects? | N | Y | Y | N | N | N | N | N | N | N | | N | Y | | Y |
| Advection/Dispersion Model: B=Box or Slab; G=Gaussian; K=K/numerical | G | G | G | B | B;G | K | B;G | G | B;G | B;G | | K | K | | B;G |
| Along-wind dispersion? | Y | Y | Y | Y | N | Y | Y | N | Y | Y | | Y | Y | | Y |
| Vertical wind shear? | N | Y | N | Y | Y | Y | Y | N | Y | | | Y | Y | | Y |
| Chemical reactions in plume? | Y | Y | Y | N | N | Y | N | N | N | N | | N | N | | N |
| Dry or wet deposition? | Y | Y | N | N | N | N | N | D | N | N | | N | N | | N |
| Concentration fluctuations? | N | N | N | N | N | N | N | N | N | N | | Y | N | | N |
| Number of chemicals (I=input) | I | 50,I | 83 | 5,I | | 5 | | 17 | 21 | | | 18,I | 1 | | I |
| Output | | | | | | | | | | | | | | | |
| Averaging time (minutes)(I=input) | | 15 | 1-10 | 3 | | 10 | | | | | | 0 | | | |
| Distance limits (km) | | | | | | | | 20 | 100 | | | 70 | 10 | | |
| Evaluation? | Y | N | Y | Y | Y | Y | Y | N | Y | Y | Y | Y | Y | N | N |
| How are data presented T=Table; G=Graph | | T;G | T;G | | T | T | T;G | T;G | T;G | T | T;G | T;G | T;G | T;G | T;G |

Table A-2 (page 2)

| | HEAVYPUFF | HEGADAS | INPUFF2.0 | MIDAS | ModSys | PLUME PATH | RIMPUFF | SAFEMODS | SAFER/TRACE | SLAB | SPILLS | SRI PUFF | TRAUMA | VAPID |
|---|---|---|---|---|---|---|---|---|---|---|---|---|---|---|
| **Operating Information** | | | | | | | | | | | | | | |
| Form of model: H=Hardware; S=Software | S | S | S | S;H | S | S | S | S;H | S;H | S | S | S | S | S |
| Main use: R=Research; A=Applied | R | R;A | A | R;A | A | R;A | A | R;A | A | R;A | A | R | A | R |
| Operate in interactive mode? | N | N | N | Y | N | N | Y | Y | Y | N | Y | Y | N | N |
| Support system? | N | N | Y | Y | N | N | Y | | Y | | | N | N | |
| Number sold or given away? | | 1 | 120 | 28 | | >10 | 3 | | 75 | | >100 | | | |
| Link to emergency system? | | N | N | Y | N | N | | | Y | | | | N | |
| **Input Data** | | | | | | | | | | | | | | |
| Accept real time weather data? | N | N | N | Y | N | N | Y | Y | Y | N | N | N | N | N |
| Method of data entry: H=Hand; F=data file memory; D=disk or tape | H;F;D | F | D | H;F;D | H;F | F | F | H;F;D | H;F;D | F | H | H;F;D | F | H;F;D |
| Source Emissions Model? | N | Y | N | Y | Y | N | Y | Y | Y | Y | Y | N | Y | Y |
| Evaporation of Spilled Liquids? | | Y | | Y | N | | | Y | Y | Y | Y | | Y | Y |
| Flashing | | N | | Y | N | | | Y | Y | N | Y | | Y | Y |
| Multicomponents | | N | | Y | Y | | Y | Y | Y | N | N | | N | Y |
| Entrainment as aerosols? | | N | | N | N | | | Y | Y | N | N | | N | Y |
| Heat transfer, substrate to cloud? | | Y | | Y | N | | | Y | Y | Y | Y | | N | Y |
| Number of substrates (W=Water; S=Soil) | | W;S | | W;S | | | | W;S | W;S | | S | | | |
| Mass transfer in liquid phase | | N | | Y | N | | | Y | Y | N | Y | | Y | Y |
| Evaporation of aerosols | | N | | N | N | | | Y | Y | N | N | | T | Y |
| Gas flux from container rupture? | | N | | Y | Y | | Y | Y | Y | N | Y | | Y | Y |
| Condens. of moisture in vapor cloud? | | Y | | N | N | | | Y | Y | N | N | | Y | Y |
| Wind influence on evaporation? | | N | | Y | N | | | Y | Y | N | Y | | N | Y |
| Number of chemicals hardwired (I=input by user) | | I | | I | 2 | | 25 | I | I | 1 | 34+ | | I | |
| Transport and Dispersion Model? | Y | Y | Y | Y | Y | Y | Y | Y | Y | Y | Y | Y | Y | Y |
| Releases treated: I=Instantaneous; C=Continuous; V=Variable | I | C;V | C;V | V | V | C | I;C;V | I;C;V | I;C;V | C | C,V | V | C | V |
| Dense cloud? | Y | Y | N | Y | Y | Y | N | Y | Y | Y | N | N | N | Y |
| Jet? | Y | N | N | Y | Y | Y | N | Y | Y | N | N | N | Y | N |
| Neutral cloud? | Y | N | Y | Y | Y | Y | Y | Y | Y | N | N | Y | N | Y |
| Buoyant cloud? | Y | N | Y | Y | Y | Y | Y | Y | N | N | N | N | N | N |
| Surface roughness? | Y | Y | N | Y | Y | N | Y | Y | Y | Y | N | N | N | Y |
| Complex terrain handled? | N | N | N | Y | N | N | Y | Y | Y | N | N | Y | N | N |
| U variations in time and space? | N | N | Y | Y | N | N | Y | Y | Y | N | N | Y | N | N |
| Indoor Concentrations? | N | N | N | Y | N | N | N | Y | Y | N | N | N | N | N |
| Building wake effects? | N | N | N | Y | N | N | N | N | N | N | N | N | N | N |
| Advection/Dispersion Model: B=Box or Slab; G=Gaussian; K=K/numerical | B | B;G | G | B;G;K | G;K | G | G | B;G | B;G | B | G | G | B | B |
| Along-wind dispersion? | Y | Y | N | Y | Y | N | Y | Y | Y | N | N | Y | N | N |
| Vertical wind shear? | Y | Y | N | Y | Y | N | Y | Y | N | N | Y | N | N | Y |
| Chemical reactions in plume? | N | N | N | N | Y | N | N | Y | Y | N | N | N | Y | N |
| Dry or wet deposition? | N | N | Y | Y | N | N | Y | Y | Y | N | N | N | Y | D |
| Concentration fluctuations? | N | N | N | N | Y | N | Y | Y | Y | N | N | N | N | N |
| Number of chemicals (I=Input) | 4 | I | | I | 2 | I | 25 | 900 | I | 1 | 34+ | | 1 | 1 |
| **Output** | | | | | | | | | | | | | | |
| Averaging time (minutes)(I=input) | | I | 3,60 | | | 0 | | 0.5,1 | | I | 10 | 60 | | 10 |
| Distance limits (km) | | | 100 | 10 | | 1 | 20 | 5 | | | 10 | | | 10 |
| Evaluation? | N | Y | Y | Y | Y | | Y | Y | Y | Y | Y | | N | N |
| How are data presented T=Table; G=Graph | T;G | T | T | T;G | T;G | T | T;G | T;G | T;G | T;G | T,G | | T | T;G |

142

Table A-2 (page 3)

| | SAFE | SAFETI | TECJET | WHAZAN |
|---|---|---|---|---|
| Operating Information | | | | |
| Form of model: H=Hardware; S=Software | S | S | S | S |
| Main use: R=Research; A=Applied | A | A | A | A |
| Operate in interactive mode? | Y | Y | Y | Y |
| Support system? | Y | Y | Y | Y |
| Number sold or given away? | 1 | 7 | 0 | 60 |
| Link to emergency system? | N | N | N | N |
| | | | | |
| Input Data | | | | |
| Accept real time weather data? | Y | N | N | N |
| Method of data entry: H=Hand; | H;F;D | H;F;D | H;F | H |
| F=data file memory; D=disk or tape | | | | |
| | | | | |
| Source Emissions Model? | N | Y | Y | Y |
| Evaporation of Spilled Liquids? | | Y | N | Y |
| Flashing | | Y | Y | Y |
| Multicomponents | | N | N | N |
| Entrainment as aerosols? | | Y | Y | Y |
| Heat transfer, substrate to cloud? | | Y | N | Y |
| Number of substrates (W=Water; S=Soil) | | S | | S |
| Mass transfer in liquid phase | | Y | N | Y |
| Evaporation of aerosols | | Y | Y | Y |
| Gas flux from container rupture? | | Y | Y | Y |
| Condens. of moisture in vapor cloud? | | Y | Y | Y |
| Wind influence on evaporation? | | N | N | Y |
| Number of chemicals hardwired | | | | |
| (I=input by user) | | 60,I | 62,I | 20,I |
| | | | | |
| Transport and Dispersion Model? | Y | Y | Y | Y |
| Releases treated: I=Instantaneous; | | | | |
| C=Continuous; V=Variable | C;V | I;C;V | C | I;C;V |
| Dense cloud? | N | Y | Y | Y |
| Jet? | Y | Y | Y | Y |
| Neutral cloud? | Y | Y | Y | Y |
| Buoyant cloud? | Y | Y | Y | Y |
| Surface roughness? | Y | Y | Y | Y |
| Complex terrain handled? | N | N | N | N |
| U variations in time and space? | Y | N | N | N |
| Indoor Concentrations? | N | N | N | N |
| Building wake effects? | N | N | N | N |
| Advection/Dispersion Model: B=Box | | | | |
| or Slab; G=Gaussian; K=K/numerical | G;K | B;G | | B;G |
| Along-wind dispersion? | Y | N | N | N |
| Vertical wind shear? | Y | N | N | N |
| Chemical reactions in plume? | Y | N | N | N |
| Dry or wet deposition? | Y | Y | N | N |
| Concentration fluctuations? | Y | N | N | N |
| Number of chemicals (I=Input) | I | 60,I | 62,I | 20,I |
| | | | | |
| Output | | | | |
| Averaging time (minutes)(I=input) | 60 | 10 | 10 | 10 |
| Distance limits (km) | 50 | 20 | | |
| Evaluation? | N | N | Y | N |
| How are data presented | | | | |
| T=Table; G=Graph | T;G | T,G | T,G | T,G |

143

# *Appendix B*

# Examples of Hazardous Chemical Release Scenarios

It is useful for the plant engineer to be familiar with the types of hazardous chemical release scenarios that he will be faced with.  The members of the Vapor Cloud Committee of the Center for Chemical Process Safety (CCPS) of the AIChE represent a cross-section of experience in this field, and they were asked to prepare a few examples of hypothetical hazardous chemical release scenarios.  Their contributions were briefly reviewed in Section 2 and summarized in Table 2-1.  The examples cover pipeline and tank ruptures, liquid spills, and runaway reactions.  Modeling results are not given for these examples.  Detailed discussions of these scenarios are given below.

B.1  General Description of a Vapor Cloud Dispersion Study

   Study Objective

   The objective of a typical study is to determine the maximum downwind and crosswind penetrations of toxic or flammable vapor clouds resulting from the accidental release of hazardous materials into the atmosphere.  For flammable releases, the hydrocarbon content of the flammable cloud between the flammable limits is also calculated as a function of downwind distance and time.  This entails source rate modeling and dense gas dispersion modeling.

   Materials Studied

   a.    Pure components ($H_2$, $NH_3$, $SO_2$, $SO_3$, $Cl_2$, HF, $SCL_2$, $C_3$, $C_4$, etc.).
   b.    Multi-components (gasoline, unstabilized naphtha, etc.).
   c.    Flashing liquids with and without aerosol formation.
   d.    Non-flashing liquids.
   e.    Vapor releases from pipelines.

   Release Scenarios Evaluated

   Most studies involve examining Largest Practicable and Largest Potential releases.  A Largest Practicable release is one with some reasonable potential of occurrence.  A Largest Potential release assumes catastrophic rupture of either the storage vessel or large diameter piping.  Some typical examples are:

**Appendix B**

<u>Largest Practicable</u>

+ rupture of small bore piping, 1" maximum
+ partial flange gasket blow-out of large diameter piping (e.g.
  50% blowout of a 2" line resulting in an equivalent hole diameter
  of 1")
+ failure of a 3/4" fusible plug on a 1 ton cylinder
+ generally limited release duration (15 minutes typical based on
  time required for operator intervention to stop the leak)

<u>Largest Potential</u>

+ rupture of 2" or 3" liquid line
+ tank truck rupture on a highway (3" or 4" assumed hole size
  typical)
+ typically, entire source vessel inventory spilled

<u>Toxic Concentration Modeled</u>

Penetration of toxic vapor clouds are typically calculated to two or three
concentration levels:

<u>Level 1</u>

Concentration at or below which most healthy people would suffer no
more severe nor long-lasting effects than extreme discomfort.

<u>Level 2</u>

Concentration above which a substantial fraction of those exposed
could be seriously injured or killed.

<u>Level 3</u>

Concentration that is immediately dangerous to life and health
(IDLH).

These concentration limits should be based on results from
toxicologic research on laboratory animals as well as past human
experience with the specific toxic compounds studied.  Typically, some
percentage of the NIOSH STEL's and TLV's are used.

<u>Spill Surface</u>

a. concrete, b. insulated concrete, c. earth/gravel, d. water, e. diked
or undiked.

## B.2  Releases of Specific Chemicals

### B.2.1  Hypothetical Ammonia Release

#### Problem Statement

Determine maximum downwind and crosswind penetrations of toxic vapor clouds resulting from an accidental release of anhydrous ammonia from a pressurized storage vessel. Penetrations to be calculated to concentrations of 100 vppm and 1,000 vppm for the Largest Practical and Largest Potential release scenarios.

#### Source Vessel

1.  40.0 metric tonnes anhydrous ammonia
2.  7.5 bar (g) pressure
3.  20 deg. C. temperature
4.  3.8 m diameter; 7.6 m initial liquid height

#### Release Scenarios

##### Largest Practical Release:

1.  rupture of 1.25" liquid line
2.  release point about 50 m from the tank
3.  two-phase choked flow assumed = 7.5 kg/s
4.  spill duration of 15 minutes
5.  undiked spill on concrete (20 deg. C)

##### Largest Potential Release:

1.  rupture of a 3" unloading line
2.  release point about 50 m from the tank
3.  two-phase choked flow assumed : 44.2 kg/s
4.  entire inventory spilled
5.  undiked spill on concrete (20 deg. C)

#### Ammonia Characterization

1.  18.8 weight % flashes upon release
2.  Assumed aerosol content of 0.35 kg liquid entrained per kg ammonia vapor (required to maintain a cloud density greater than that of air over an ammonia concentration range of 0.0 to 1.0 mole fraction).

Another example of an ammonia release would be the case of a tank holding cryogenic ammonia being overfilled from a barge. The liquid is discharged from the overflow pipe at the liquid transfer rate.  The material flashes and the plume moves downwind.

B.2.2  Hypothetical Chlorine Release

Chlorine is transported and stored in one ton cylinders or "bullets".  These cylinders are of heavy steel, about 3 feet in diameter and about 6 feet long.  Chlorine is dispensed for water treatment and other uses by means of small diameter steel tubing from the vapor space of the cylinder.  A release to the atmosphere can occur in two ways:

1.      The cylinder is cracked open; most likely dropped from the bed of the delivery truck.

2.      The tubing is somehow opened or broken off, thus releasing a jet.

In the first case, the liquid contents will flash to form a cloud of dense gas and a liquid pool.  The cloud will immediately travel downwind.  The pool will start to evaporate at a rate which depends upon the temperature and thermal conductivity of the ground as well as the wind speed and turbulence conditions.  To model this, the contents of the cylinder can be adiabatically flashed to atmospheric pressure to find the temperature and amounts of liquid and vapor.  A boiling or non-boiling evaporation model, depending upon the ground conditions, can then be used to calculate evaporation rates from a given area which can then be input to an area source dispersion model (heavy gas or neutral buoyancy, depending upon the circumstances) for estimation of downwind chlorine concentrations.  This scheme is not considered further here.

The second case, breakage of the dispensing tube, will be considered in more detail.  The saturation pressure of chlorine at most ambient temperatures is at least twice atmospheric.  If the tube is broken off completely, choked (or critical) flow of the vapor will occur.  The maximum flow that could occur would be that through a sharp-edged orifice with the same diameter as the tube.  (These tubes have inside diameters on the order of 0.2 to 0.3 inches.)  Thus the maximum flow rate can be found by calculating the choked flow rate of chlorine vapor at the temperature and pressure of the bullet contents, generally ambient conditions.  The bulk temperature can be assumed to be 100 degrees Fahrenheit for which the saturation vapor pressure is about 157 psia.  The equations provide the mass flux, from which the mass flow rate can be calculated using the area of the throat, and then the thrust (which is often of interest).  The density of the fluid in the throat is assumed to correspond to an ideal gas temperature of 25 degrees F.

The fluid will exit the orifice at the throat pressure and immediately expand to atmospheric pressure.  As a first approximation, this step in the formation of the jet can be modeled by assuming that the fluid expands at the throat velocity to atmospheric pressure, instantaneously.  This implies that there is no time for a significant amount of air to be entrained.  Thus to calculate the expanded jet diameter, the expanded jet initial density is required.  This can be done by an adiabatic flash calculation or graphically by using a Mollier diagram for the material.  However, for these approximate purposes, it can be assumed that the temperature change for the adiabatic expansion is small (say, less than $5^O$ F) so the expanded jet density can be calculated for an ideal gas at $25^OF$ and 14.7 psia.


## Specific Chlorine Scenarios


A facility consists of a 1 ton cylinder of chlorine with a 1" feed line (Schedule 80).  Determine the hazard zones that could result from:

1. Continuous release of chlorine caused by popping the safety plug on the cylinder (1/2" diameter).  Plug releases at $158^OF$.  Assume liquid release.

2. Instantaneous release of one ton of chlorine caused by tank rupture under fire conditions with failure of the plug to open.  Tank design pressure is 500 psig (DOT 106-A500).

3. Continuous release of chlorine caused by a break in the 1" line and failure of all devices to isolate the leak from the tank at ambient temperature of $77^OF$.

     Assume a 10' line length
     Assume a 100' line length

4. External Fire - A tank of chlorine is involved in a pool fire situation.  A LPG pipe line ruptures and the liquid fraction of the butane pools under the chlorine tank.  The vapor is ignited and results in a pool fire impinging on the chlorine tank.  The pool size is one inch deep and 10 ft. in diameter.

B.2.3  Hypothetical Release of TMS from a Safety Relief Valve

The problem is to estimate the maximum concentration at several air intakes during a hypothetical release of trimethoxysilane (TMS) from a safety relief valve.  TMS reacts almost instantaneously with moisture in the air to produce methanol.  The IDLH concentration for TMS (47 ppm) is used as a guideline concentration criterion as the duration of the release is estimated as 33 minutes.

The high velocity release (V=206 fps) of TMS vapor is directed vertically from a release height of 42 feet.  The TMS (MW=122) is four times more dense than air (MW=29).  The jet reaches the peak of its trajectory above the relief valve and then falls to the ground.  The flow and emission rates are estimated from the safety valve rating.  The release duration is then estimated from the tank volume.

The "worst case" scenario occurs for the windspeed and direction which directs the trajectory of the jet near the receptor.  The frequency of occurrence of the "worst case" windspeed and direction are not of interest.  It is further assumed that the ambient temperature is 0 degrees Centigrade and that the TMS does not react due to the low ambient moisture content.  The release is assumed to be all vapor.

B.2.4  Hypothetical Release of Ethylene Oxide

There is concern about the ambient ethylene oxide concentration in the vicinity of an ethylene glycol unit.  An engineer samples the emissions from a evaporator vent and determines the ethylene oxide emission rate.  He also measures the vent diameter, flow rate, and height.  The vent emissions and ground level ethylene oxide concentrations must be modeled.

The conditions modeled at a specific ethylene glycol unit are a 41.0 lb/hr ethylene oxide emission rate, a vent height of 50 feet, a vent diameter of 4.0 inches, a total flow rate of 500 acfm, an exhaust temperature of $229^{O}F$, and an ambient temperature of $80^{O}C$.  The emission is considered to be continuous for modeling purposes.

The TLV concentration for ethylene oxide is 1.0 ppm.  The TLV is a time-weighted average concentration for a normal 8 hour work day and a 40 hour work week, to which nearly all workers could be repeatedly exposed, day after day, without adverse effect.

B.2.5  Short-Term Release of Hydrogen Sulfide from a Still

Hydrogen sulfide would be released if there is a failure of the hydrogen sulfide still or pipe network connected to the still.  The plant engineering department is interested in determining ground level concentrations for the following hypothetical release scenario.

Approximately 160 lbs. of hydrogen sulfide will be released in 1 minute if the still is vented to atmospheric pressure.  If another 10 minutes is required to stop the gas flow then another 280 lbs will be released.  The temperature of the hydrogen sulfide is $55^{O}C$ and the ambient temperature is $25^{O}C$ and the release height is 85 feet.  A total of 440 pounds of hydrogen sulfide is released in 11 minutes.

The Immediate Dangerous to Life and Health (IDLH) concentration for hydrogen sulfide is 300 ppm.  The IDLH concentration is the maximum ambient

concentration to which a person may be exposed for 30 minutes without suffering
health damage.  The threshold odor limit for hydrogen sulfide is 0.0005 ppm.

B.2.6  Hypothetical Case of Emulsion Kettle Runaway Reaction with
       Dense Two-Phase Release

A potential problem is the flammable/explosion hazard associated with a
dense two-phase (vapor with entrained liquid) release resulting from the
emergency venting of a reactor.  Research has indicated that some of our
emulsion polymerization systems will foam while venting to form a homogeneous
mixture of vapor and liquid.  Since a substantial amount of liquid could be
entrained with the vapor as an aerosol, the resulting two-phase mixture leaving
the reactor is dense and has the potential of falling toward the ground where
ignition sources are present.

The event is assumed to take place in an existing reactor.  The scenario
assumes that all batch charges are made to the reactor, then the contents
undergo an exothermic reaction which increases the temperature and pressure in
the vessel until the set pressure of the rupture disk is reached.  The disk
breaks and the reactor contents are vented as a two-phase mixture through the
rupture disk vent line.  To simplify the case, it can be assumed that the vapor
is all butadiene and that the liquid phase is all water.

The 7,000 gallon reactor is located inside a building.  The existing
rupture disk system is 600 mm in diameter with a 150 mm vent line discharging
7.5 meters above the building roof.  The vent line length is 17.5 meters.  The
emission point elevation is 25 meters above ground level.

Two release profiles should be calculated for the events, based on Design
Institute for Emergency Relief System (DIERS) technology, assuming that the
vessel behavior and the vent line flow is homogeneous (i.e. the mass fraction
of liquid entering the vent line is the same as the mass fraction of liquid in
the vessel).  Release parameters are desired over the duration of the release.
Source velocity, flow rate, jet pseudo diameter (after expansion to atmospheric
pressure), plume density, vapor concentration, and water concentration should
be calculated for various elapsed times.

A lower flammable limit of 2.00 volume percent should be used for the
vapor mixture.  The maximum ground level concentrations are of interest.
Maximum predicted ground level concentrations and flammable cloud size should
be calculated for several meteorological cases.

B.2.7  Hypothetical Case of $H_2S$ Pipeline Rupture

A 12-inch-diameter buried pipeline carries natural gas with a 10% (by
mass) concentration of $H_2S$ from a well to a gas treatment plant two kilometers
away.  The conditions of gas in the pipeline are:  Temperature = 300K, Pressure
= 1,000 psig, and Molecular Weight of Gas = 19.  Assume that the pipeline is

damaged by a backhoe at a location midway between the well and gas plant such that the line is completely severed in half, and gas is released horizontally. Determine the following:

1.  Mass release rate of $H_2S$ with time.

2.  Initial dilution of gas cloud caused by violent expansion of the pressurized gas.

3.  Distance from release point at which momentum jet effects dominate.

4.  Height of gas cloud above the ground when and if buoyancy effects dominate.

5.  Maximum 1-minute peak and 15-minute average concentrations of $H_2S$ gas at 1,000 meters downwind assuming Stability Class F and 1 m/s wind speed.

B.2.8   HCH Release

A maintenance man is tightening the bolts on a valve bonnet following a job to tighten the valve stem packing.  The nuts are overtightened causing the bonnet of the valve to pop off and lose the full flow due to 12 ft. of liquid head from a nearby tank.

HCH at 54°F

B.2.9   Acrolein Release

A hose breaks during filling of an acrolein car.  The hose diameter is 2" with a flow rate of 200 gpm.  There is no dike in the area where the hose break takes place.

# References

REFERENCES CITED IN TEXT

Air Weather Service, 1978:  Calculation of toxic corridors, AWS Pamphlet 105-57.

Ale, B. and R. Whitehouse, 1986: Computer based system for risk analysis of process plant. Heavy Gas and Risk Assessment-III, S. Hartwig (ed.), D. Reidel, Dordrecht.

Alp, E., 1985:  COBRA:  An LNG Model. Heavy Gas Workshop - Toronto. Proc. produced by Concord Scientific Corporation, Downsview.

Alp, E., S. Alp, P.M. Brown and R.V. Portelli, 1985:  Heavy gas dispersion modeling.  Paper VI.2, Proc. of 15th Int. Tech. Meeting on Air Poll. Modeling and its Applic., NATO/CCMS.

Balentine, H.W., and M.W. Eltgroth, 1985:  Validation of a hazardous spill model using $N_2O_4$ and LNG spill data.  Paper No. 85-25B.1, APCA Annual Meeting.

Bell, R.P., 1978:  Isopleth calculations for ruptures in sour gas pipelines. Energy Processing/Canada, 36-39.

Blackmore, D.R., M.N. Herman and J.L. Woodward, 1982:  Heavy gas dispersion models.  J. Haz. Mat., 6, 107-128.

Blewitt, D.N., 1986:  Computerized modeling rupture design analysis for sour gas pipeline safety analysis.  Standard Oil Co. (Indiana), 200 E. Randolph Dr., Chicago, IL 60601.

Bodurtha, F.T., 1961:  The behavior of dense stack gases.  J. Air Poll. Control Assoc., 11, 431-435.

Briggs, G.A., 1969:  Plume Rise, AEC Critical Review Series, Report TID-24635.

Briggs, G.A., 1975:  Plume Rise Predictions, in Lectures on Air Pollution and Envir. Impact Analysis, AMS, 59-111.

Briggs, G.A., 1984:  Plume Rise Buoyancy Effects, Atmospheric Science and Power Production, DOE/TIC-27601, 327-366.

Brighton, P.W., A.J. Prince and D.M. Webber, 1985:  Determination of cloud area and path from visual and concentration records.  J. Haz. Mat., 11, 155-178.

**References**

Britter, R., 1979:  The spread of a negatively buoyant plume in a calm
    environment.  Atmos. Environ., 13. 1241-1247.

Britter, R., 1980:  The ground-level extent of a negatively buoyant plume in a
    turbulent boundary layer.  Atmos. Environ., 14, 779-785.

Britter, R.E. and R.F. Griffiths, eds., 1982:  Dense Gas Dispersion,
    Elsevier, New York, 247 pp..

Chan, S.T., 1983:  FEM3-A finite element model for the simulation of heavy gas
    dispersion and incompressible flow users manual.  UCRL-53397,LLL,
    Livermore, CA.

Chan, S.T., P.M. Gresho, D.L. Ermak, 1981:  A Three-Dimensional, Conservation
    Equation Model for Simulating LNG Vapor Dispersion in the Atmosphere.
    UCID-19210, Lawrence Livermore Lab., Livermore, CA.

Chatwin, P.C., 1982:  The use of statistics in describing and predicting the
    effects of dispersing gas clouds.  Dense Gas Dispersion, (ed. by Britter
    and Griffiths), Elsevier, New York, 213-230.

Chatwin, P.C., 1983:  The incorporation of wind shear effects into box models
    of heavy gas dispersion.  Atmos. Disp. of Heavy Gases and Small Particles
    (eds. G. Ooms and H. Tennekes) Springer-Verlag, 63-72.

Clewell, H.J., 1983:  A simple formula for estimating source strengths from
    spills of toxic liquids, ESL-TR-83-03.

Colenbrander, G.W., 1980:  A mathematical model for the transient behavior of
    dense vapor clouds.  Proc., 3rd Int. Symp. on Loss Prevention and Safety
    Promotion in the Process Industries.

Colenbrander, G.W. and J.S. Puttock, 1983:  Maplin Sands experiments 1980:
    Interpretation and modeling of liquified gas spills onto the sea.  Atmos.
    Disp. of Heavy Gases and Small Particles, IUTAM Symp. (eds, G. Ooms, H.
    Tennekes), Springer-Verlag.

Cox, R.A. and R.J. Carpenter, 1979:  Further developments of a dense vapor
    cloud dispersion model for hazard analysis.  Symp. on Heavy Gas
    Dispersion, Frankfurt, published in Heavy Gas and Risk Assessment (ed. S.
    Hartwig), D. Reidel, Dordrecht.

Cox, W.M. and J.A. Tikvart, 1985:  Assessing the performance level of air
    quality models.  Paper V. 3 at 15th Int. Tech. Meeting on Air Pollution
    Modeling and its Applications.  NATO/CCMS.

Crabol, B., A. Roux and V. Lhomme, 1986:  Interpretation of the Thorney Island Phase 1 Trials with the Box Model CIGALE2.  Second Symp. on Analysis and Interpretation of Results of the Thorney Island Trials.

Deaves, D.M., 1984:  Application of advanced turbulence models in determining the structure and dispersion of heavy gas clouds.  Atmos. Disp. of Heavy Gases and Small Particles, IUTAM Symp. (eds. G. Ooms, H. Tennekes) Springer-Verlag.

Deaves, D.M., 1985:  3-dimensional model predictions for the upwind building trial of Thorney Island Phase II.  J. Haz. Mat., 11, 341-346.

deNevers, N., 1984:  Spread and down-slope flow of negatively-buoyant clouds. Atmos. Environ., 18, 2023-2028.

Drivas, P.J., 1982:  Calculation of evaporative emissions from multicomponent liquid spills.  Environ. Sci. Tech., 16, 726-728.

Drivas, P.J., J.S. Sabnis and L.H. Teuscher, 1983:  Model simulates pipeline, storage tank failures.  Oil and Gas Journal, Sept. 12, 1983, 162-169.

Duijm, N.J., A.P. van Ulden, W.H. van Heugten and P.J. Builtjes, 1985: Physical and mathematical modeling of heavy gas dispersion - accuracy and reliability.  Paper VI.3, Proc. of 15th Int. Tech. Meeting on Air Poll. Modeling and its Applic., NATO/CCMS.

Efron, B., 1982:  The jacknife, the bootstrap and other resampling plans. CBMMS-NSF-38.  Soc. Ind. and Appl. Math., Philadelphia, 92 pp.

Eidsvik, K.J., 1980:  A model for heavy gas dispersion in the atmosphere. Atmos. Environ., 14, 769-777.

Eltgroth, M. and C. Smith, 1983:  CHARM$^{TM}$ Emergency Response System Documentation.  RADIAN Corp., P.O. Box 9948, Austin, TX 78766.

England, W.G., L.H. Teuscher, L.E. Hauser and B. Freeman, 1978:  Atmospheric dispersion of liquefied natural gas vapor clouds using SIGMET, a three-dimensional time-dependent hydrodynamic computer model.  1978  Heat Transfer and Fluid Mechanics Institute.

Environmental Protection Agency, 1984:  Interim Procedures for Evaluating Air Quality Models (Revised), EPA-450/4-84-023.  QAQPS, EPA, Research Triangle Park, NC.

Environmental Protection Agency, 1984:  Evaluation and Selection of Models for Estimating Air Emissions from Hazardous Waste Treatment, Storage, and Disposal Facilities, EPA-450/3-84-020, OAQPS, EPA, Research Triangle Park, NC.

# References

Ermak, D.L. and S.T. Chan, 1985:  A study of heavy gas effects on the atmospheric dispersion of dense gases.  Paper VI.6, Proc. of 15th Int. Tech. Meeting on Air Poll. Modeling and its Applic., NATO/CCMS.

Ermak, D.L., S.T. Chan, D.L. Morgan and L.K. Morris, 1982:  A comparison of dense-gas dispersion model simulations with Burro series LNG spill test results.  J. Haz. Mat., 6, 129-160.

Fauske, H.K., 1985:  Flashing flows or:  Some practical guidelines for emergency releases.  Plant/Operations Progress, 4, 132-134.

Fay, J.A., 1986:  The dispersion of dense gases in the atmosphere.  Proc. AMS Short Course on Atmos. Disp.

Fay, J.A. and D. Ranck, 1981:  Scale effects in liquified fuel vapor dispersion.  DOE/EP-0032, USDOE, Washington, DC.

Fay, J.A. and D.A. Ranck, 1983:  Comparison of experiments on dense gas cloud dispersion.  Atmos. Environ., 17, 239-248.

Fay, J.A. and S.G. Zemba, 1986:  Integral model of dense gas plume dispersion. Atmos. Environ., 20, 1347-1354.

Feldbauer, G.F. et al, 1972:  Spills of LNG onto water: vaporization and downwind drift of combustible mixture.  Esso Res. and Eng. Co., Report EE61E-72.

Fleischer, M.T., 1980: SPILLS, An evaporation/air dispersion model for chemical spills onland.  Shell Devel. Center, Westhollow Res. Center, P.O. Box 1380, Houston, Texas 77001.

Fryer, L.S., 1980:  CRUNCH.  A Computer program for the calculation of the dispersion of continuous releases of denser-than-air vapors. HSE/SRD/PD101/WP5, Safety and Reliability Directorate, Culcheth, UK.

Fryer, L.S. and G.D. Kaiser, 1979:  DENZ - A computer program for the calculation of the dispersion of dense toxic or explosive gases in the atmosphere, SRD R 152 UKAEA, Culcheth, UK.

Germeles, A.E. and E.M. Drake, 1975:  Gravity spreading and atmospheric dispersion of LNG vapor clouds.  Proc., 4th Int'l. Symp. on Transport of Hazardous Cargoes by Sea and Inland Waterways.  Jacksonville, FL, 519-539.

Gifford, F.A., 1976:  Turbulence diffusion typing schemes - a review. Nuc. Safety, 17, 68-86.

Goldwire, H.C., T.G. McRae, G.W. Johnson, D.L. Hipple, R.P. Koopman, J.W. McClure, L.K. Morris and R.T. Cederwall, 1985: Desert Tortoise Series Data Report, 1983 Pressurized Ammonia Spills. UCID-20562, Lawrence Livermore Nat. Lab, Livermore, CA.

Hanna, S.R., 1982: Applications in Air Pollution Modeling. Atmospheric Turbulence and Air Pollution Modeling (eds. F.T.M. Nieuwstadt and H. van Dop., D. Reidel, Dordrecht, 275-310.

Hanna, S.R., 1984: The exponential pdf and concentration fluctuations in smoke plumes. Bound. Lay. Meteorol., 29, 361-376.
Hanna, S.R., 1986: A review of air quality model evaluation procedures. WMO Int. Conf. on Air Poll. Modeling and its Applications, Leningrad, USSR.

Hanna, S.R., G.A. Briggs and R.P. Hosker, 1982: Handbook on Atmospheric Diffusion. DOE/TIC-11223, DOE, 102 pp.

Hanna, S.R. J.C. Weil and R.J. Paine, 1986: Plume Model Development and Evaluation: Hybrid Approach. Final Report, EPRI, 3412 Hillview Avenue, Palo Alto, CA 94303.

Hart, G.S., 1986: Avoiding and managing environmental damage from major industrial accidents. J. Air Poll. Control Assoc., 36, 127-138.

Havens, J.A., 1978: An assessment of the predictability of LNG vapor dispersion from catastrophic spills onto water. Proc., 5th Int. Symp. on the Transport of Dangerous Goods by Sea and Inland Waterways.

Havens, J.A. and T.O. Spicer, 1985: Development of an Atmospheric Dispersion Model for Heavier-than-Air Gas Mixtures. Report No. CG-D-22-85 for the U.S. Coast Guard by the Chem. Eng. Dept., University of Arkansas, Fayetteville, Ark. 72701, Three Volumes.

Heinold, D.W., R.J. Paine, K.C. Walker and D.G. Smith, 1986: Evaluation of the AIRTOX dispersion algorithms using data from heavy gas field experiments. Proc. 5th Conf. on Application of Air Poll. Meteorol., AMS, Boston.

Hoot, T.G., R.N. Meroney and J.A. Peterka, 1973: Wind tunnel tests of negatively buoyant plumes, CER73-74TGH-RNM-JAP-13, Colorado State Univ., Fort Collins.

Horst, T.W. and J.C. Doran, 1986: Nocturnal drainage flow on simple slopes. Boundary Layer Meteorology, 34, 263-286.

Humbert-Basset, R. and A. Montet, 1972: Dispersion dans l'atmosphere d'un nuage gaseux forme par epandage de GNL sur le sol, Proc. 3rd Int. Conf. on LNG, Paper VI.4

# References

Ille, G. and C. Springer, 1978:  The evaporation and dispersion of hydrazine propellants from ground spills, CEEDO-TR-78-30, AD A059407.

Jagger, S.F., 1983:  Development of CRUNCH:  A dispersion model for continuous releases of a denser-than-air vapor into the atmosphere. SRD R 229, UKAEA, Culcheth, UK.

Jensen, N.O., 1983:  On cryogenic liquid pool evaporation. J. Haz. Mat., 8, 157-163.

Kaiser, G.D., and B.C. Walker, 1978:  Releases of anhydrous ammonia from pressurized containers - the importance of denser-than-air mixtures. Atmos. Environ., 12, 2289-2300.

Kayes, P.J. (ed.), 1985: Manual of industrial hazard assessment techniques. World Bank, London, England.

Kelty, J., 1984:  Calculation of evacuation distances during toxic air pollution incidents. Atmospheric Emergencies:  Existing Capabilities and Future Needs, Transp. Res. Record 902.  National Academy of Science, Washington, DC.

Koopman, R.P., R.T. Cederwall, D.L. Ermak, H.C. Goldwire, W.J. Hogan, J.W. McClure, T.G. McRae, P.O. Morgan, H.C. Rodean, and J.H. Shinn, 1982: Analysis of Burro series $40m^3$ LNG spill experiments. J. Haz. Mat., 6, 43-83.

Koopman, R.P., T. G. McRae, H.C. Goldwire, D.L. Ermak and E.J. Kansa, 1984: Results of recent large-scale $NH_3$ and $N_2O_4$ dispersion experiments.  UCRL-91830, 3rd Symp. on Heavy Gases and Risk Assessment.

Kunkel, B.A., 1983:  A comparison of evaporative source strength models for toxic chemical spills, Air Force Geophysics Laboratory, AFGL-TR-83-0307, Hanscom AFB, MA 01731.

Kunkel, B.A., 1985:  Development of an atmospheric diffusion model for toxic chemical releases, Air Force Geophysics Laboratory, AFGL-TR-85-0338.

Layland, D.E., D.J. McNaughton and P.M. Bodner, 1986:  Assessing the reliability of dispersion models for hazardous materials releases. Proc. 5th Joint Conf. on Applic. of Air Poll. Meteorol., AMS.

Leahey, D.M. and M.M. Schroeder, 1985:  Predictions of maximum ground-level $H_2S$ concentrations resulting from two sour gas well blowouts.  Western Research.

Leung, J.C., 1986:  A generalized correlation for one-component homogeneous equilibrium flashing choked flow. AIChE Journal, 32, 1743-1746.

Lewellen, W.S. and R.I. Sykes, 1986:  Analysis of concentration fluctuations from lidar observations of atmospheric plumes.  J. Clim. and Applied Meteorol., 25, 1145-1154.

Lewellen, W.S., R.I. Sykes and S.F. Parker, 1985:  Comparison of the 1981 INEL dispersion data with results from a number of different models. NUREG/CR-4159 NRC, Washington, DC.

Londergan, R.J., D.H. Minott, D.J. Wackter and R.R. Fizz, 1983:  Evaluation of Urban Air Quality Simulation Models, EPA-450/4-83-020.

Ludwig, F.L., 1985:  Users' guide for a model of airflow and diffusion in complex terrain (MADICT), prepared for U.S. Army Research Office by SRI Int., Menlo Park, CA.

Mackay, D. and R.S. Matsugu, 1983:  Evaporation rates of liquid hydrocarbon spills on land and water, Can. J. Chem. Eng., 51, 434-439.

McRae, T.G., 1985:  Evaluation of source strength and dispersion model predictions with data from large nitrogen tetroxide field experiments. Paper VI.4, Proc. of 15th Int. Tech. Meeting on Air Poll. Modeling and its Applic., NATO/CCMS.

McRae, T.G., R.T. Cederwall, H.C. Goldwire, Jr., D.L. Hipple, G.W. Johnson, R.P. Koopman, J.W. McClure, and L.K. Morris, 1983:  Eagle Series Data Report:  1983 Nitrogen Tetroxide Spills, Lawrence Livermore National Laboratory Report UCID-20063.

McCready, D.I., J.R. Foster and C.E. Grigsby, 1986:  A BASIC dispersion model for toxic and hazardous gas releases.  Paper 86-93.3 Proc., 79th Ann. Meeting of APCA.

McNaughton, D.J., M.A. Atwater, P.M. Bodner, and G.G. Worley, 1986:  Evaluation and Assessment of Models for Emergency Response Planning.  Prepared for CMA by TRC, 800 Connecticut Blvd. East Hartford, CT 06108.

McQuaid, J, 1985:  Objectives and design of the Phase I heavy gas dispersion trials.  J. Haz. Mat., 11, 1-34.

Meroney, R.N., 1979:  Lift off of buoyant gas initially on the ground.  J. Ind. Aerodyn., 5, 1-11.

Meroney, R.N., 1982:  Wind-tunnel experiments on dense gas dispersion.  J. Haz. Mat., 6, 85-106.

Meroney, R.N., 1984:  Transient characteristics of dense gas dispersion.  Part II: Numerical experiments on dense cloud physics.  J. Haz. Mat., 9, 159-170.

Meroney, R.N. and A. Lohmeyer, 1984:  Prediction of propane cloud dispersion by a wind-tunnel data calibrated box model.  J. Haz. Mat., 8, 205-221.

**References**

Meroney, R.N. and D.E. Neff, 1984:  Wind tunnel simulation of cold gas spills. Proc. 1984 ASME Ann. Meeting.

Mikkelsen, T., S.E. Larsen and S. Thykier - Nielsen, 1984:  Description of the Riso Puff Diffusion Model. Nuclear Technology, 67, 56-65.

Ministry of the Environment, 1983:  Portable computing system for use in toxic gas emergencies. Report No. ARB-162-83-ARSP, MOE, Toronto, Canada.

Ministry of the Environment, 1986:  Review and Modification of Heavy Gas Model. MOE, Toronto, Canada.

Morgan, D.L., L.K. Morris and D.L. Ermak, 1983:  SLAB:  A time-dependent computer model for the dispersion of heavy gases released in the atmosphere.  UCRL-53383, Lawrence Livermore National Laboratory, Livermore, CA.

Oke, T.R., 1978:  Boundary Layer Climates, John Wiley and Sons, New York.

Oliverio, M., E. Alp, C.P. Bourque and C.S. Matthias, 1986:  COBRA III Enhancements.  Prepared for Environment Canada by Concord Scientific Corporation, Toronto, Canada.

Ooms, G., 1972:  A new method for the calculation of the plume path of gases emitted by a stack.  Atmos. Environ., 6, 899-909.

Ooms, G. and N.J. Duijm, 1984:  Dispersion of a stack plume heavier than air. Atmos. Disp. of Heavy Gases and Small Particles, IUTAM Symp. (eds, G. Ooms, H. Tennekes), Springer-Verlag, 1-23.

Ooms, G., A.P. Mahieu, and F. Zelis, 1974:  The plume path of vent gases heavier than air.  First Int. Symp. on Loss Prevention and Safety Promotion in the Process Industries.  The Hague.

Ooms, G. and H. Tennekes, eds., 1984:  Atmospheric Dispersion of Heavy Gases and Small Particles.  Springer-Verlag, Berlin Heidelberg.

Paine, R.J., J.E. Pleim, D.W. Heinold and B.A. Egan, 1986:  Physical processes in the release and dispersion of toxic air contaminants.  Paper 86-76.3 at 79th Ann. Meeting of the Air Poll. Control Assoc., Minneapolis.

Perry, R.H., D.W. Green, and J.O. Maloney, 1984:  Perry's Chemical Engineer's Handbook, Sixth Edition, McGraw-Hill, New York. (Note:  This Handbook is Revised on a regular basis.)

Petersen, W.B. and L.G. Lavdas, 1986:  INPUFF 2.0 - A Multiple Source Gaussian Puff Dispersion Algorithm.  Users Guide, ASRL, USEPA, Research Triangle Park, NC.

Picknett, R.G., 1981:  Dispersion of dense gas puffs released in the atmosphere at ground level.  Atmos. Environ., 15, 509-525.

Pielke, R., 1981:  Mesoscale Numerical Modeling.  Adv. in Geophysics, 23, 185-344.

Puttock, J.S., 1985:  Thorney Island data and dispersion modeling.  J. Haz. Mat., 11, 381-398.

Puttock, J.S., D.R. Blackmore and G.W. Colenbrander, 1982:  Field experiments on dense gas dispersion.  J. Haz. Mat., 6, 13-41.

Puttock, J.S. and G.W. Colenbrander, 1985:  Dense-gas dispersion-experimental research.  Heavy Gas Workshop - Toronto, Proc. produced by Concord Scientific Corporation, Downsview, Ontario, Canada.

Puttock, J.S. and G.W. Colenbrander, 1985:  Thorney Island data and dispersion modeling.  J. Haz. Mat., 11, 381-397.

Puttock, J.S., G.W. Colenbrander and D.R. Blackmore, 1982:  Maplin Sands Experiments 1980:  Dispersion Results from Continuous Releases of Refrigerated Liquid Propane.  Proc. Heavy Gas and Risk Assessment - II (S. Hartwig, ed.) D. Reidel, Dordrecht, 147-161.

Puttock, J.S., G.W. Colenbrander and D.R. Blackmore, 1984:  Maplin Sands 1980: Dispersion results from continuous releases of refrigerated liquid propane and LNG.  Air Poll. Modeling and its Applic. (C. DeWispelaere, ed.), Plenum, 353-373.

RADIAN Corporation, 1986:  Description of the Radian Complex Hazardous Air Release Model (CHARM).  Version 4.0, P.O. Box 9948, Austin, TX 78766.

Raj, P.K., 1981:  Models for cryogenic liquid spill behavior on land and water. J. Haz. Mat., 5, 111-130.

Raj, P.K., 1985:  Summary of heavy gas spills modeling research.  Proc. of Heavy Gas (LNG/LPG) Workshop.  Toronto, 51-75.

Riou, Y.J. and A.E. Saab, 1985: A three-dimensional numerical model for the dispersion of heavy gases over complex terrain.  Paper VI.7, Proc. of 15th Int. Tech. Meeting on Air Poll. Modeling and its Applic., NATO/CCMS.

Scire, J.S. and F. Lurmann, 1983:  Development of the MESOPUFF II Dispersion model.  Proc., Sixth Symp. on Turbulence and Diffusion, AMS, 32-35.

Shaw, P. and F. Briscoe, 1978:  Evaporation from spills of hazardous liquids on land and water.  SRD R 100 UKAEA, Culcheth, UK.

**References**

Spicer, T.O, J.A. Havens, P.A. Tebean, and L.E. Key, 1986:  DEGADIS - A heavier-than-air gas atmospheric dispersion model developed for the U.S. Coast Guard.  Paper 86-42.2, Proceedings, APCA Ann. Conf., Minneapolis.

Stiver, W. and D. MacKay, 1983:  Evaporation rates of chemical spills. Environment Canada First Technical Chemical Spills Seminar, Toronto.

Strimaitis, D.G. and W.H. Snyder, 1986:  An evaluation of the complex terrain dispersion model against laboratory observations:  neutral flow over 2-D and 3-D hills.  Proc., Fifth Joint Conf. on Applic. of Air Poll. Meteorol., AMS 215-218.

Sutton, O.G., 1953:  Micrometeorology, McGraw-Hill, New York.

Swift, I., 1984:  Development in emergency relief system design. The Chem. Eng., Aug./Sept., 30-33.

van Ulden, A.P., 1974:  On the spreading of a heavy gas released near the ground.  Proc. Loss Prevention and Safety Promotion in the Process Industries, 221-226.

van Ulden, A.P., 1983:  A new bulk model for dense gas dispersion: two-dimensional spread in still air.  Atm. Disp. of Heavy Gases and Small Particles.  IUTAM Symp. (eds. G. Ooms and H. Tennekes), 419-440.

Verholek, M.G., 1986:  CARE-Modeling hazardous airborne releases.  Proc., Nat. Conf. on Hazardous Wastes and Hazardous Materials.

Webber, D.M., 1983:  The physics of heavy gas cloud dispersal, SRD R243, UKAEA, Culcheth, UK.

Webber, D.M. and P.W. Brighton, 1986:  A mathematical model of a spreading, vaporizing liquid pool.  Heavy Gas and Risk Assessment III, (ed. S. Hartwig), D. Reidel.

Wheatley, C.J., P.W.M. Brighton and A.J. Prince, 1985:  Comparison between data from the heavy gas dispersion experiments at Thorney Island and predictions of simple models.  Paper VI.5, Proc. of 15th Int. Tech. Meeting on Air Poll. Modeling and its Applic., NATO/CCMS. Wu,

Whitacre, C.G., J.H. Griner III, M.M. Myirski and D.W. Sloop, 1986:  Personal Computer Program for Chemical Hazard Prediction (D2PC).  U.S. Army Chem. Res. and Develop. Center, Aberdeen Proving Ground, MD 21010.

Wilson, D.J., 1979:  The release and dispersion of gas from pipeline ruptures. Prepared for Alberta Environment.

Wilson, D.J., 1981:  Along-wind diffusion of source transients.  Atmos. Environ., 15, 489-495.

Wu, J.M. and J.M. Schroy, 1979:  Emissions from Spills, APCA Spec. Conf. on Control of Specific (Toxic) Pollutants, Pittsburgh.

Xiao-Yun, L., H. Leijdens and G. Ooms, 1986:  An experimental verification of a theoretical model for the dispersion of a stack plume heavier than air. Atmos. Environ., 20, 1087-1094.

Yamada, T., 1985:  Numerical simulations of valley ventilation and pollutant transport.  Proc., Seventh Symp. on Turbulence and Diffusion, AMS, 312-314.

Zannetti, P., G. Carboni, R. Lewis and L. Matamala, 1986:  AVACTA II - Users Guide (Release 3.1), AV-R-86/530, AeroVironment, Inc., Monrovia, CA 91016.

Zannetti, P., 1986:  A new mixed segment-puff approach for dispersion modeling. Atmos. Environ., 20, 1121-1130.

Zeman, O., 1982:  The dynamics and modeling of heavier-than-air, cold gas releases.  Atmos. Environ., 16, 741-752.

**References**

ADDITIONAL REFERENCES NOT CITED IN TEXT


Anderson, G.E., H. Hogo and G.Z. Whitten, 1984:  Development and Application
    of Methods for Estimating Effects of Industrial Emission Controls on Air
    Quality Impacts of Reactive Pollutants.  SYSAPP-84/116, SAI, 101 Lucas
    Valley Road, San Rafael, CA 94903.


ApSimon, H.M. and A.C. Davison, 1986:  A statistical model for deriving
    probability distributions of contamination for accidental releases.
    Atmos. Environ., 20, 1249-1260.


Astleford, W.J., T.B. Morrow and J.C. Buckingham, 1986:  Hazardous Chemical
    Vapor Handbook for Marine Vessels.  Report CG-D-12-83, U.S. Coast Guard,
    Washington, DC.


Bjorklund, J.R. and R.K. Dumbauld, 1981:  User's Instructions for the Volume
    Source Diffusion Models Program and the Volume/Line Source Graphs Computer
    Programs, Tech. Rep. TR-81-321-08, by H.E. Cramer Co., Inc., prepared for
    U.S. Army Dugway Proving Ground.


Brighton, P.W., 1985:  Evaporation from a plane liquid surface into a turbulent
    boundary layer.  J. Fluid Mech., 159, 323-345.


Brighton, P.W.M., 1986:  Heavy-gas dispersion from sources inside buildings
    or in their wakes.  I. Chem. E. Symp. on Refinement of Estimates of the
    Consequences of Heavy Toxic Vapour Releases.


Britter, R.E., 1982:  Special topics on dispersion of dense gases.  RPG 1200,
    prepared for HSE, Sheffield, UK.


Byggstoyl, S. and L.R. Saetran, 1983:  An integral model for gravity spreading
    of heavy gas clouds.  Atmos. Environ., 17, 1615-1621.


Carn, K.K. and P.C. Chatwin, 1985:  Variability and heavy gas dispersion.
    J. Haz. Mat., 11, 281-300.


Carn, K.K., S.J. Sherrell and P.C. Chatwin, 1986:  Analysis of Thorney Island
    data:  variability and box models.  Stably Stratified Flow and Dense Gas
    Dispersion to be published by Oxford Univ. Press (ed. J.S. Puttock).


Chatwin, P.C. and P.J. Sullivan, 1986:  Note on intermittency and probability
    density functions of scalars in turbulence.  To appear in J. Fluid Mech.


Cheah, S.C., S.K. Chua, J.W. Cleaver, and A. Millward, 1983:  Modeling of
    heavy gas plumes in a water channel.  Atmos. Disp. of Heavy Gases and
    Small Particles, (eds. G. Ooms and H. Tennekes) Springer-Verlag, 199-210.

Cheah, S.C., J.W. Cleaver, and A. Millward, 1983:  The physical modeling of heavy gas plumes in a water channel.  Proc., 4th Int. Symp. on Loss Prevention and Safety Promotion in the Process Industries, Harrogate, England.

Chong, J., 1980, GAS.  Ministry of the Environment, Toronto, Canada.

Clary, J.C. and P. Shetty, 1986:  SPILLS for IBM-PC.  Trinity Consultants, Richardson, TX.

Colenbrander, G.W. and J.S. Puttock, 1983:  Dense gas dispersion behavior: experimental observations and model developments.  Proc. 4th Int. Symp. on Loss Prevention and Safety Promotion in the Process Industries.

Connell, J.R. and H.W. Church, 1980:  Recommended methods for estimating atmospheric concentrations of hazardous vapors after accidental release near nuclear reactor sites.  NUREG/CR-1152, Sandia Labs, Albuquerque, NM.

Csanady, G.T., 1969:  Dosage probabilities and area coverage from instantaneous point sources on ground level.  Atmos. Environ., 3, 25-46.

Davies, A.E. and V. Ferraro, 1986:  Using Physical Scale Models to Identify and Remedy Stack Gas and Spill Dispersion Problems.  Paper 86-42.6 at APCA Conf.

Deaves, D.M., 1986:  Development and application of heavy gas dispersion models of varying complexity.  Proc. 2nd Symp. on Heavy Gas Disp. Trials at Thorney Island.

Dirkmaat, J.J., 1981:  Experimental studies on the dispersion of heavy gases in the vicinity of structures and buildings.  Proc. Symp. Designing with the Wind, Nantes.

Dumbauld, R.K., J.R. Bjorklund and J.F. Bowers 1970:  Handbook for Estimating Toxic Fuel Hazards.  CR-61326, NASA.

Eidsvik, K.J, 1981:  Heavy gas dispersion model with liquified release.  Atmos. Environ., 15, 1163-1164.

Eidsvik, K.J., 1981:  Prediction of hazard area of an accidental gas release.  Proc. Seventh Conf. on Prob. and Stat. in Atmos. Sci., AMS, 86-89.

Evans, A. and J.S. Puttock, 1986:  Experiments on the ignition of dense flammable gas clouds.  Proc. 5th Int. Symp. on Loss Prevention and Safety Promotion in the Process Industries.

**References**


Fabrick, A.J., 1982:  CHARM:  An operational model for predicting the fate
    of elevated or surface releases of dense or buoyant gases.  Preprints, 3rd
    Joint Conf. on Applic. of Air Poll. Meteorol., AMS.

Fay, J.A., 1980:  Gravitational spread and dilution of heavy vapor clouds.
    Proc. 2nd Int. Symp. on Strat. Flows, 471-494.

Fields, D.E., C.W. Miller and S.J. Cotter, 1984:  Validation of the AIRDOSE-EPA
    computer code by simulating intermediate range transport of $^{85}$Kr from the
    Savannah River Plant.  Atmos. Environ., 18, 2029-2036.

Goldwire, H.C., T.G. McRae, G.W. Johnson, D.L. Hipple, R.P. Koopman, J.W.
    McClure, L.K. Morris and R.T. Cederwall, 1983:  Coyote series data report,
    LLNL/NWC 1981 LNG spill tests, dispersion, vapour burn, and rapid phase
    transition.  UCID-19953, LLNL, Livermore, CA.

Gotaas, Y., 1985:  Heavy gas dispersion and environmental conditions as
    revealed by the Thorney Island experiments.  J. Haz. Mat., 11, 399-408.

Griffiths, R.F. and A.S. Harper, 1985:  A speculation on the importance of
    concentration fluctuations in the estimation of toxic response to irritant
    gases.  J. Haz. Mat., 11, 369-372.

Griffiths, R.F. and L.C. Megson, 1984:  The effects of uncertainties in human
    toxic response on hazard range estimation for ammonia and chlorine.
    Atmos. Environ., 18, 1195-1206.

Gugan, K., 1979:  Unconfined Vapor Cloud Explosions, Inst. of Chem. Eng. in
    assoc. with George Godwin Ltd.

Hall, D.J., E.J. Hollis and H. Ishaq, 1984:  A wind tunnel model of the Poston
    dense gas spill fluid trials.  Atm. Disp. of Heavy Gases and Small Part.
    IUTAM Symp., Springer-Verlag, Berlin, 189-198.

Hall, D.J. and R.A. Waters, 1985:  Wind tunnel model comparisons with the
    Thorney Island dense gas release field trials.  J. Haz. Mat., 11, 209-236.

Handbook for Chemical Hazard Predictions, 1977:  U.S. Army Material
    Development and Readiness Command, Arlington, VA.

Hanna, S.R., 1986:  Spectra of concentration fluctuations - the two time
    scales of a meandering plume.  Atmos. Environ., 20, 1131-1138.

Hanna, S.R. and B. Munger, 1983:  A survey of emergency models and applications
    - Part I:  Models for hazardous spills applications.  Paper 83-26.1, APCA
    Annual Meeting.

Hansen, F.V., 1979:  Engineering Estimates for the Calculation of
    Atmospheric Dispersion Coefficients.  ASL, White Sands Missile Range, NM.

Hanzevack, E.L., 1982:  Dispersion from safety valves and other momentum
     emission sources.  Atmos. Environ., 16, 1231-1242.

Hartwig, S., 1985:  Improved understanding of heavy gas dispersion due to the
     analysis of the Thorney Island trials data.  J. Haz. Mat., 11, 417-423.

Havens, J.A., 1979:  A description and assessment of the SIGMET LNG vapor
     dispersion model.  Rep. CG-M-3-79, U.S. Coast Guard, Washington.

Havens, J.A., 1982:  MIT-GRI LNG Safety and Research Workshop.  Vol. II
     Dispersion of Dense Vapors, GRI 82/0019.2.

Jensen, N.O., 1981:  Entrainment through the top of a heavy gas cloud.
     Air Poll. Modeling and its Applic. I (ed. C. DeWispelaere), Plenum Press,
     477-487.

Jensen, N.O., 1984:  Entrainment through the top of a heavy gas cloud,
     numerical treatment.  Air Poll. Modeling and Its Appl. III
     (ed. C. De Wispelaere), Plenum Press, 343-351.

Johnson, M.C., 1980:  Methodology for Chemical Hazard Prediction.  AD-A086820,
     NTIS, Springfield, VA.

Kahler, J.P., R.G. Curry and R.A. Kandler, 1980:  Calculating Toxic Corriders,
     AWS/TR-80/003, Air Weather Service (MAC), Scott AFB, IL.

Krogstad, P.A. and R.M. Pettersen, 1986:  Wind tunnel modeling of a release of
     a heavy gas near a building.  Atmos. Environ., 20, 867-878.

Leahey, D.M. and M.J.E. Davies, 1984:  Observations of plume rise from sour gas
     flares.  Atmos. Environ., 18, 917-922.

Leggett, R.V., 1979:  A Mathametical Model for Non-Uniform Simple Surface
     Evaporation of a Liquid Contaminant (NUSSE), ARCSL-TM-79012, Chemical
     Systems Lab., Aberdeen Proving Ground, MD.

Li, W.W. and R.N. Meroney, 1983:  Gas dispersion near a cubical model building.
     J. Wind Eng. and Ind. Aerodyn., 12, 15-47.

Ludwig, F.L., 1981:  A model for simulating the behavior of pollutants
     emitted at ground level under time-varying meteorological conditions.
     Atmos. Environ., 15, 989-1000.

Ludwig, F.L., 1984:  Transport and diffusion calculations during time- and
     space-varying meteorological conditions using a microcomputer.  J. Air
     Poll. Control Assoc., 34, 227-232.

## References

Ludwig, F.L., 1986:  A nonsteady-state adaptive volume plume model suitable for regions of strong vertical gradients of wind and stability.  Proc., 5th Conf. on Applic. of Air Poll. Meteorol., AMS.

McQuaid, J., 1986:  Some effects of obstructions on the dispersion of heavy gas clouds.  Symp. on Refinement of Estimates of the Consequences of Heavy Toxic Vapor Releases.

Meroney, 1983:  Gas dispersion near a cubical model building. J. Wind Eng. and Ind. Aerodyn., 12, 15-47.

Meroney, R.N., 1986:  Guideline for fluid modeling of liquified natural gas cloud dispersion.  Report No. GRI-86/0102.1, prepared for GRI, Chicago, IL 60631.

Misra, P.K., 1985:  Importance of concentration fluctuations in LNG and heavy gas modeling. Heavy Gas Workshop - Toronto, proc. produced by Concord Scientific Corp., Downsview.

Morrow, T.B., 1982:  Development of an improved pipeline breakflow model, SwRI-6917/3, Am. Petrol. Inst., Washington, DC.

Morrow, T.B., J.C. Buckingham and F.T. Dodge, 1984:  Water channel tests of dense plume dispersion in a turbulent boundary layer. Atmos. Disp. of Heavy Gases and Small Particles. IUTAM Symp. (eds. G. Ooms and H. Tennekes), 323-331.

Napier, D., M. Oliverio and E. Alp, 1986:  Source terms for dispersion modeling of releases of heavy gases.  Presented at the Montreal Seminar for Chemical Spills.

Neff, D.E. and Meroney, R.N., 1982:  The behavior of LNG vapor clouds:  Wind tunnel test on the modeling of heavy plume dispersion.  GRI/0145, Gas Research Institute, Chicago.

Netterville, D.D.J., 1986:  Plume rise and dispersion in turbulent winds:  an application for acoustic remote sensing, to appear in Atmos. Environ.

Nussey, C., J.K.W. Davies and A. Mercer, 1985:  The effect of averaging time on the statistical properties of sensor records. J. Haz. Mat., 11, 209-236.

Ohmstede, W.D., R.K. Dumbauld and G.G. Worley, 1983:  Ocean Breeze/Dry Gulch Equation Review.  ESL-TR-83-05, Air Force Engineering and Services Center, Tyndall AFB, FL.

Ontario Ministry of the Environment, 1983:  Portable computing system for use in toxic gas emergencies, Report No. ARB-162-83-ARSP.

Prince, A.J., 1985:  Details and results of spill experiments of cryogenic liquids onto land and water.  SRD R 324, UKAEA, Culcheth, UK.

Puttock, J.S., 1986:  Dispersion of dense vapor clouds in contact with the ground-theory and experiment.  Proc. IMA Conf. on Math. in Major Accident Risk Assess. Oxford Un. Press.

Puttock, J.S., 1986:  Comparison of Thorney Island data with predictions of HEGABOX/HEGADIS.  Symp. on Anal. and Interpret. of Results of Thorney Island Trials, to appear in J. Haz. Mat.

RADIAN Corporation, 1985:  Emergency modeling of hazardous air pollutants.  ESL-TR 84-55, prepared for AFESC, Tyndall AFB, FL 32403.

Raj, P.K., 1986:  Dispersion of hazardous reactive chemical vapors in the atmosphere.  Proc., 1986 Hazardous Material Spills Conference, 305-314.

Raj, P.K. and R.C. Reid, 1978:  Fate of liquid ammonia spilled onto water. Env. Sci. and Tech., 12, 1422-1425.

Reijnhart, R., J. Piepers, R. Rose and L.H. Toneman, 1980:  Vapor cloud dispersion and the evaporation of volatile liquids in atmospheric wind fields.  Atmos. Environ., 14, 751-762.

Ride, D.J., 1984:  A probabilistic model for dosage.  Atm. Disp. of Heavy Gases and Small Particles.  IUTAM Symp. (eds. G. Ooms and H. Tennekes), 267-276.

Rottman, J.W., J.C.R. Hunt, and A. Mercer, 1985:  The initial and gravity-spreading phases of heavy gas dispersion:  comparison of models with Phase I data.  J. Haz. Mat., 11, 261-280.

Saucier, R., 1981:  A Mathematical model for the vapor and mass from a falling, evaporating aerosol cloud.  ARCSL-TR-81007, Chemical Systems Lab., Aberdeen Proving Ground, MD.

Sebacher, D.I., W.R. Cofer, III, D.C. Woods and G.L. Maddrea, 1984:  Hydrogen chloride and aerosol ground cloud characteristics resulting from Space Shuttle launches.  Atmos. Environ., 18, 763-770.

Spicer, T.O. and J.A. Havens, 1985:  Modeling the Phase I Thorney Island experiments.  J. Haz. Mat., 11, 237-260.

Stern, E., 1986:  The Portable Computing System for Use in Gas Emergencies. Risk Analysis, 6, 381-385.

Webber, D.M., 1984:  Gravity spreading in dense gas dispersion models.  Atmos. Disp. of Heavy Gases and Small Particles, IUTAM Symp.

**References**

Webber, D.M., 1986:  Evaporation and boiling of liquid pools- a unified treatment.  Proc. IMA Conf. on Math. in Major Risk Assess.

Webber, D.M. and C.J. Wheatly, 1986:  The effect of initial potential energy on the dilution of a heavy gas cloud.  Proc. 2nd Symp. on Heavy Gas Dispersion Trials at Thorney Island.

Wheatley, C.J., 1986:  Factors affecting cloud formation from releases of liquified gases.  I. Chem. E. Symp. on Refinement of Estimates of the Consequences of Heavy Toxic Vapour Releases.

Wilson, D.J. and B.W. Simms, 1985:  Exposure time effects on concentration fluctuations in plumes.  Dept. of Mech. Eng.  The University of Alberta, prepared for Alberta Environment.

Zeman, O., 1984:  Entrainment in gravity currents.  Atm. Disp. of Heavy Gases and Small Particles, IUTAM Symp. (eds. G. Ooms and H. Tennekes), 53-62.

# Author Index

# Subject Index